Prai

"The Inside the Minds series
and techniques of accomplis
Thelen Reid & Priest

"Unlike any other business book." - Bruce Keller, Partner, Debevoise & Plimpton

"A must read for anyone in the industry." - Dr. Chuck Lucier, Chief Growth Officer, Booz-Allen & Hamilton

"A snapshot of everything you need..." - Charles Koob, Co-Head of Litigation Department, Simpson Thacher & Bartlet

"A great way to see across the changing marketing landscape at a time of significant innovation." - David Kenny, Chairman & CEO, Digitas

"An incredible resource of information to help you develop outside-the-box..." - Rich Jernstedt, CEO, Golin/Harris International

"Tremendous insights..." - James Quinn, Litigation Chair, Weil Gotshal & Manges

"Great information for both novices and experts." - Patrick Ennis, Partner, ARCH Venture Partners

"A rare peek behind the curtains and into the minds of the industry's best." - Brandon Baum, Partner, Cooley Godward

"Unique insights into the way the experts think and the lessons they've learned from experience." - MT Rainey, Co-CEO, Young & Rubicam/Rainey Kelly Campbell Roalfe

"Intensely personal, practical advice from seasoned dealmakers." - Mary Ann Jorgenson, Coordinator of Business Practice Area, Squire, Sanders & Dempsey

"Great practical advice and thoughtful insights." - Mark Gruhin, Partner, Schmeltzer, Aptaker & Shepard, P.C.

Inside the Minds

The critically acclaimed *Inside the Minds* series provides readers of all levels with proven business intelligence from C-level executives (CEO, CFO, CTO, CMO, partner) from the world's most respected companies. Each chapter is comparable to a white paper or essay and is a future-oriented look at where an industry/profession/topic is heading and the most important issues for future success. Each author has been carefully chosen through an exhaustive selection process by the *Inside the Minds* editorial board to write a chapter for this book. *Inside the Minds* was conceived in order to give readers actual insights into the leading minds of business executives worldwide. Because so few books or other publications are actually written by executives in industry, *Inside the Minds* presents an unprecedented look at various industries and professions never before available.

INSIDE THE MINDS

Winning Legal Strategies for Biotech Companies

Intellectual Property Protection, FDA Approval, and Board Management in the Biotech Industry

BOOK IDEA SUBMISSIONS

If you are a C-level executive or senior lawyer interested in submitting a book idea or manuscript to the Aspatore editorial board, please e-mail authors@aspatore.com. Aspatore is especially looking for highly specific book ideas that would have a direct financial impact on behalf of a reader. Completed books can range from 20 to 2,000 pages—the topic and "need to read" aspect of the material are most important, not the length. Include your book idea, biography, and any additional pertinent information.

SPEAKER SUBMISSIONS FOR CONFERENCES

If you are interested at giving a speech for an upcoming ReedLogic conference (a partner of Aspatore Books), please e-mail the ReedLogic Speaker Board at speakers@reedlogic.com. If selected, speeches are given over the phone and recorded (no travel necessary). Due to the busy schedules and travel implications for executives, ReedLogic produces each conference on CD-ROM, then distributes the conference to bookstores and executives who register for the conference. The finished CD-ROM includes the speaker picture with the audio of the speech playing in the background, similar to a radio address played on television.

INTERACTIVE SOFTWARE SUBMISSIONS

If you have an idea for an interactive business or software legal program, please e-mail software@reedlogic.com. ReedLogic is especially looking for Excel spreadsheet models and PowerPoint presentations that help business professionals and lawyers achieve specific tasks. If idea or program is accepted, product is distributed to bookstores nationwide.

Published by Aspatore, Inc.
For corrections, company/title updates, comments, or any other inquiries, please e-mail store@aspatore.com.

First Printing, 2005
10 9 8 7 6 5 4 3 2 1

ISBN 1-59622-294-8

Library of Congress Control Number: 2005931819

Inside the Minds Managing Editor, Leah M. Jones, Edited by Michaela Falls, Proofread by Eddie Fournier

Winning Legal Strategies for Biotech Companies

Intellectual Property Protection, FDA Approval, and Board Management in the Biotech Industry

CONTENTS

Legal Strategies for Biotechnology Companies

Laura A. Coruzzi, Ph.D.
Partner
Jones Day

The First Meeting with a Client

Different biotechnology companies have different legal problems—problems that are unique to whatever marketplace they are involved in. There is no "one size fits all."

In any initial meeting with a client, I spend a lot of time listening not only to the science, but to the business and business goals of the client. Both are important. The stage of the company is another factor. If they are at an early stage, then you are engaging in what we call "science projection"—trying to predict where this technology is going to be ten years from now, and even further out. When a company is in late-stage clinical trials and about to launch product, they have a more immediate set of needs. Their cash is probably going into the clinical trial, and they are usually focused on that one product. Depending on what stage they are at and what their marketplace is, the advice varies. In the first meeting, you have to get a sense of the technology and the business plan. My job is to figure out a patent strategy that will adequately protect their technology from unauthorized use by competitors and identify potential barriers (i.e., freedom to operate issues the company may face).

Areas of Difficulty for Clients

Unless a client is very sophisticated in patent law, they often misconstrue the rights conferred by a patent. It is quite common to think that having a patent gives you the right to do something, but a patent does not give you the right to do anything in the active sense. It gives you the right to exclude others from making, using, selling, offering for sale, and importing the invention. This stems from Article I, Section 8 of the Constitution, which says that Congress shall have power to reward inventors for their works by giving them, for limited times, the exclusive right to their inventions. Congress interpreted "exclusive" to mean the right to exclude others from making, using, selling, offering for sale, and importing. Many clients think, "I have a patent on that—that means I can do it." In fact, you have to check third-party patents to make certain there are no dominating rights or patents covering components or your method of manufacturing that could affect your freedom to operate. Time and time again you hear, "But we have a patent on that." It does not matter, because it does not shield your activities.

Legal Issues Specific to Biotechnology

From a legal standpoint, the pharmaceutical/biotechnology industry is different from other industries in some significant ways because of the nature of the inventions and the high degree of regulation of medicines. For example, the patent statute contains an "FDA-exemption"—a specific statutory exemption from infringement for clinical trial work related to obtaining FDA approval. This exemption (part of the Hatch-Waxman Act) can seriously affect your analysis of third-party patents and freedom to operate issues. A pharmaceutical client may be able to take advantage of this statutory exemption from infringement. Depending on the life left of the third-party patents, it might work to your client's benefit. On the other hand, the exemption applies to your client's patents as well, and can "weaken your hand." Right now, the industry is waiting for a Supreme Court decision on whether this exemption applies to pre-clinical trial work. It is unclear what impact that decision will have moving forward.

For biologics, the constant struggle from a patent point of view is that many of the products are naturally occurring proteins, genes, antibodies, and the like. From the beginning of biotechnology patent law in the early '80s, companies have been faced with the challenge of describing how the hand of man changed this natural substance so that it becomes potentially patentable.

On another front, in many other industries, the doctrine of inherency is relatively rarely applied, but in our area it is a lively one. For example, courts have invalidated claims to particular chemical entities referred to as "metabolites" (drugs are converted to other chemical forms, called metabolites, by the body). Metabolite patents have been invalidated based on prior use of the parent drug that was converted in patients' bodies to the metabolite. These issues affect companies trying to extend patent life in order to recapture time in regulatory that ate away at the commercially relevant period of patent protection. You have to be vigilant and clever about how you straddle these issues.

Protecting Intellectual Property

Again, protecting intellectual property is not a one-size-fits-all process for biotechnology companies. Usually, when a startup company comes to us with a new technology or a new therapeutic approach, they are generally focused on one disease or one molecular entity. We try to work with the client to explore ways in which its protection can be justifiably expanded. An example is the work we handled for a client that developed an antiviral treatment for HIV/AIDS. When they came to us, they had discovered segments of proteins in the HIV virus that, if synthesized as peptides, inhibited the infectivity of the virus. By way of background, to be infectious, the virus has to fuse to the cell to gain entry and do its damage. The antiviral peptides interfered with fusion—they were "anti-fusogenic." When we started discussing this mechanism, as the scientists were explaining it, it seemed to us to be perhaps an underlying mechanism that many other enveloped viruses that cause disease might use to gain entry into cells. So we asked if this could be applied to these other viruses. The answer was "probably"; it seemed to be a basic mechanism that many viruses used. The next question was how we would dope out the peptides that would be applicable to those other viruses.

As a result of the discussions, the clients developed a computer program that could identify cognate sequences in the other viruses. When we filed their patent applications, we covered the gamut—not just HIV, but every other virus that caused serious infectious disease that had the cognate peptide. The portfolio that ensued covered the treatment of HIV, but it also covered cognate peptides and treatments of the other infectious diseases. The patent portfolio staked out much more commercial territory than the company had originally envisioned entering. In general, the parallel patent rights can be parlayed into licenses that could be granted to other companies to develop products in those areas. In other words, you can use the patents to generate revenue in areas where you know you do not have the wherewithal to be commercially active. The patent right could become an income generator for you that otherwise would have been lost to another company, who would have filed on it once they discovered those cognate drugs.

Policing the Portfolio

Biotechnology companies must police their portfolios to make certain they are earning the appropriate fees, and that no one else is overstepping their patents. Many biotechnology companies have in-house counsel that have different ways of searching the literature and trade publications for announcements. The industry knows itself. While it is very hard to figure out in-house activities of the competition, such as what methodologies they may be using or what screening activities may be going on, eventually, because the scientists live in the "publish-or-perish" world, it has to come out. Even if your competitors are filing their own patent applications, usually the scientific journals are a source of what they are up to. Eventually, the clinical trials get published as well.

One of our clients had a situation in which their competitor was monitoring what they were doing in clinical trials and filing patent applications covering everything our client was developing commercially. We managed to bring it to a head by an unusual legal proceeding in which we actually sued the patent office under the Administrative Procedures Act to prevent the issuance of the patent to the competitor. However, we had a peculiar set of facts. We actually had a ruling from the patent office that the competitor's patents did not describe our product. On the one hand, one arm of the patent office determined that our competitor's patent application was inadequate to claim our product, and on the other hand a rogue examiner was willing to allow them a claim that would do just that. We reasoned that you should not have two disparate decisions coming out of the patent office, and took it to the courts to reconcile it. The case settled favorably to our clients.

Now, there are mechanisms for companies to monitor each other, and many legal strategies can be activated to ensure you cover your competitor's materials, or, on the other side of the coin, to ensure you defend against that. The United States has moved to a European-style publication system. Before the GATT Treaty was enacted, U.S. patent applications were kept a secret unless and until they issued. Under the European-style system, U.S. applications publish eighteen months after earliest priority date claimed in the U.S. You can see what your competitors are doing in connection with the prosecution of their patents.

Best Strategies

When we are in the mode of protecting the client's crown jewels or their technology, we engage in what I refer to as "science projection." Patents are deeds to property; they are not just scientific publications. So what we try to do is engage the client in a discussion about where this technology can go commercially: what fields and what areas. Then we try to project and handle the patent portfolio by filling in the scientific details required to ensure patent rights for these future markets. I call that science projection. When we are in the defensive or offensive mode, where the client already has the patents or is up against a patent litigation, we engage more in scientific archaeology: We go back in time to when the patent was filed and recreate the scene to better defend or attack the patent. Getting experts on board early is critical. It is very important to know early on who the right players are and whom the company may need in this time of war. We have a cadre of lawyers with Ph.D.s who are rather well connected in the scientific community, and who assist us in rapping with the Nobel Prize winners and the scientists who we need to line up to help clients defend themselves.

Three Steps to Avoiding Common Legal Issues

There are three steps a biotechnology company can take to avoid common legal problems.

Step one is to spend the money to do the appropriate searching. If you have a new technology you want to protect by patents, and you want to ensure freedom to operate, identify the issues by spending the money up front, before you start investing in it. Poor planning at the beginning means paying at the back end. Clients will often say they do not want to spend a lot of money up front: "Just take a look-see." But if you do not investigate thoroughly, your whole strategy could be based on a false premise—especially if there is a prior art reference, a third-party patent, or a pending application that you missed. You can save a lot of money in the long run by doing a careful job up front. You are going to spend money to save money in the long run.

Step two relates to more mature companies. Quite often, universities or research institutes file the initial patent applications, and then companies acquire them. Although the company has no control over what the university did initially, upon acquisition the company should carry out due diligence and see if there is some way to fix what is wrong or make a course correction to better protect its business interest. Typically, research institutes do not have the same business goals that a company does, so their patent applications will be less focused on the business plan than a company's application would. The institute's tends to be broader and more far reaching. When the company acquires the right to the patent, it might need to ensure that the commercially important aspects of the technology are being adequately protected. In a sense, you have to perform triage. You cannot do everything, so you should do what I call "unscrambling the omelet"; among the jumble of patent rights you have licensed in, make sure you are prosecuting the ones that fit your business needs.

Step three is to revisit the patents often, because the company's business goals change over time. At point one, platform technology may be all important for raising money and getting the funding needed to start. Two or three years later, the company might be in a completely different place, where they have focused on a lead compound and their business goals are different. They should not be spending money on the early patents that may no longer be relevant to the business or marketplace. They have entered a new business. Triage takes place constantly; annual reviews of where the company is from a business point of view are crucial to discover any disconnects in the patent portfolio and how they can be fixed. During clinical trials, before product launch, you must start thinking of patent term extension. It is never too early to start thinking about which patents would qualify. You need to know which one would be best for you, strategically, so you get the maximum amount of protection after all the money you invested in getting the patent right. Any improvements made in the technology should trigger new rounds of patent applications, so that the second- and third-generation products receive adequate protection.

Significant Laws Affecting Biotechnology Companies

The patent statute is the same regardless of whether the company's business is biotechnology, pharmaceuticals, electronics, or mechanical devices. We are still governed by Section 101 of the patent statute, which dictates that ideas are not patentable. There are four categories of patentable subject matter: (1) process, (2) machine, (3) article of manufacture, or (4) composition of matter, and any new and useful improvements of those. I refer to that as the seat of the stool of what is patentable. Then patentability is supported by three legs. For that subject matter to be patentable, it has to be novel; this is dictated by 35 U.S.C. § 102. That means it has to be different from prior uses, sales, or publications. Even if it is novel, the second leg of the stool is that it has to be not obvious over the prior art use, sale, or publication. That is governed by Section 103. Last, but not least, is Section 112, where the quid pro quo for government giving you this monopoly is fully divulging your invention to the public. You have to have a written description that enables the skilled artisan to practice the invention without undue experimentation and set forth the best mode contemplated by the inventor for doing it. Those statutory requirements apply to every invention.

Where biotechnology differs is in the case law that has developed over time by applying these statutes to our technology. Biotechnology is unique because we are dealing with natural products and, in some cases, metabolites, so we have issues of inherency that may not be as alive in the electronics or mechanical fields. These issues pertaining to patentability/validity are further complicated by legislation affecting enforcement in the pharmaceutical/biotech arena. The statutory exemptions are driven by socioeconomic concerns to make generic drugs available to the public as soon as possible after the expiration of the patent—namely, the Hatch-Waxman Act. Here, the government struck a balance between the needs of pharmaceutical companies and companies that produce generic drugs. The law allows pharmaceutical companies to tack on additional years of patent life to compensate for the commercial time lost while obtaining FDA approval for a new drug, and generics to do clinical work prior to the expiration of the patent in order to obtain approval, so that they can market a generic equivalent the day the pharmaceutical company's patent expires. Thus, we encounter legal theories rarely used in most industries (e.g., inherency) and we have statutory exemptions to infringement that do not exist in the other industries. None of these issues seems to impact the other industries, even though the body of patent law we are dealing with is the same.

Changes in the Last Ten Years

The GATT Treaty that enacted the twenty-year rule has had a huge impact on biotechnology and pharmaceutical inventions and patents. In the electronics industry, products usually do not outlive the patents. In the pharmaceutical industry, it is very different, because so much time is spent with the FDA. Under the old law, the patent term lasted seventeen years from the date the patent issued; so the clock started ticking from the date the patent was granted. If you could time your portfolio so that your patents issued upon obtaining FDA approval, that was the perfect outcome. During the time your patent application was pending, it was an inchoate right that had huge value; however, the patent became a wasting asset the day it issued. Now the patent is a wasting asset the day it is filed, because the twenty-year clock ticks from your U.S. filing date. Now there is a rush for FDA approval, and you are dealing with a persnickety statute for adding back time that was lost during the approval process or during patent prosecution to tack on to the end of your patent life. The old seventeen-year rule was a more strategic way to operate than this mechanical set of laws. This has been a major change for companies to deal with. It is not brand new—we have been living with it since 1995 or 1996—but it has changed the way people practice.

The Supreme Court Hilton Davis decision was another major event in patent law that affects biotechnology in particular. In the early days of the biotechnology industry, the types of patents applicants sought started out with broad claims, because the technology had broad and far-reaching implications. It was common for such claims to be narrowed during examination. Now, with the Hilton Davis decision on the doctrine of equivalents, it turns out that if you start with broad claims and you are forced to narrow them during prosecution, you have effectively given up the difference in scope and could be estopped from recapturing the scope of the claims. It has changed the strategy of patent prosecution for inventions in the biotechnology industry.

Last but not least, we have a lot of Hatch-Waxman litigation going on in the drug industry. When a generic company files an Abbreviated New Drug Application (ANDA) or a paper ANDA with the FDA, it must identify third-party patents covering the drug it is imitating. If there are patents that are still in force covering the innovator's drug, the Generic has to fill out a certification (called a "Paragraph IV certification") that it either will not market its generic product until the patent expires, or that it does not infringe and/or the patent is invalid. Such certification by the Generic sets off a whole series of events. When the Paragraph IV certification is filed, it is served on the patentee or the innovator, who has forty-five days to sue the producer of the generic drug. If the innovator sues for patent infringement, an automatic thirty-month stay is put on the Generic's approval. A patent infringement case takes place during that period to see whether the generic will prevail.

In the biologics area, the FDA has not yet grappled with generics. The FDA is uncomfortable with granting the same kind of approval, because biologics are so very different from synthetic organic molecules. The agency is concerned about how you prove equivalence. For example, even if the protein sequence is the same, the host cells in which they are produced might be infected with viruses that the innovator drug company's host cells do not have. What effect does that have on the final product? Are there differences in the manufacturing process that lead to an inability to say that one batch is the same as another? There is legislation proposed for a biologic generic, but the FDA is still struggling with it. We think it is going to happen within the next few years. At that time, biologics companies will face the same type of litigation that drug companies currently face. Right now, they do not. That is something we see on the horizon.

Good Advice

One rule I learned from my now-deceased mentor, S. Leslie Misrock, who was referred to affectionately as the "father" of biotechnology patent law, is not to fall in love with the technology. Clients would come in with hot new technology, and he would tell them time and time again not to fall in love with the inventory. It was wise advice that I will never forget. You cannot get so enamored of the technology that you lose reason and business sense. You have to keep your eye on the business end, because if there are blocking patents that are insurmountable, or if patentability cannot be achieved, you are going to have a hard time raising funds. While it might be a great technology, if it cannot be protected, it won't be easy to commercialize, to say the least.

When I started out in the early '80s, pharmaceutical companies were buying biotechnology companies because they did not have that expertise in-house. Antibody companies were being bought, but big pharma was pretty much against developing protein drugs. I remember some lawyers from a big pharmaceutical company that will remain nameless saying that antibodies will never be drugs, "not on our watch." That was 1983. Now, antibodies are one of the hottest commodities out there. In the newspaper today, it appears that Genentech's antibody Avastin (bevacizumab) is promising for a new indication for small-cell lung carcinoma. Herceptin (trastuzumab, Genentech), Remicade (infliximab, Johnson & Johnson), and Humira (adalimumab, Abbott) are products that are taking off. Doctors are using them. The side effects are fewer than they are with cancer chemotherapy. While antibodies can be difficult (or more expensive) to manufacture, their exquisite sensitivity or selectivity for disease targets cannot necessarily be replicated by small organic molecules. If you make a drug that poisons a particular type of enzyme, it will poison cognates. Antibodies will only hit the cells that have the target they lock onto. Biologics, although messy to make, are exquisitely selective when used in patients. As a result, these drugs apparently have fewer side effects. I think we are going to see more and more selective drugs with fewer side effects gaining popularity in the cancer chemotherapeutic treatment area, where the patient population has tired of the standard cut, burn, and poison approach to treatment.

Globalization

The whole industry has become more global, and there are some tensions between the U.S. system and the European system. This is not peculiar to the biotechnology industry, but you need to be aware of this for pharmaceuticals and biotechnology, if not any kind of invention, because of the vagaries of the differences between the laws. As mentioned above, the Hilton Davis decision relating to the doctrine of equivalents shows us that in the U.S., it is not good to start with a broad claim and then narrow it. By contrast, in Europe, if you do not start with a broad claim, you could run afoul of Article 123. Once a patent issues in Europe, it can be opposed by any party within a nine-month period. During the opposition, the patentee is not allowed to amend their claims to go beyond what was initially issued. If you do not start out broad in Europe—if you start out narrow and that is what issues—you could get stuck during this opposition period. You will not be able to fix what is wrong, because you cannot expand the scope you began with. Conversely, in the U.S., you want to start narrower and become broader. We work very closely with our lawyers in the Munich office, who practice in the European Patent Office to make sure we do not run afoul of those issues when we are dealing with U.S. and foreign filings for our client. Strategic coordination is critical.

Laura Coruzzi worked at Pennie & Edmonds LLP from 1981 through 2003 and was one of the first members of that firm's biotechnology group founded by S. Leslie Misrock, affectionately known as the "father of biotechnology patent law." Concentrating on intellectual property issues as applied to a variety of disciplines in the biotechnological field, Dr. Coruzzi's practice has evolved with the patent laws and matured with the needs of the biotechnology/pharmaceutical industries. Her current practice involves all aspects of patent law as it relates to the emerging fields of biotechnology, including genetic engineering, molecular biology, plant molecular biology, virology, immunology, genomics, functional genomics, proteomics, diagnostics, drug discovery, pharmaceuticals, pharmacogenomics, and drug delivery.

The focus of Dr. Coruzzi's patent procurement practice centers on the strategic planning and management of patent portfolios designed to protect emerging new platform technologies (such as functional genomics, proteomics, tissue engineering, and gene therapy), biologics (antibodies, growth factors, and peptide drugs), as well as other pharmaceuticals and therapies (antivirals, anti-angiogenic therapies, and chemotherapeutics). This includes handling U.S. prosecution and interference, as well as management of corresponding foreign proceedings, particularly oppositions in the European Patent Office and Australia, and revocations and litigations in the United Kingdom. Her client counseling activities include due diligence evaluations of patent portfolios; evaluation, analysis, and opinions on the patentability of inventions; validity of patents and freedom-to-operate or clearance opinions; and involvement in licensing strategies, negotiations, and agreements.

In addition to procurement and counseling, Dr. Coruzzi's practice also encompasses litigation and an appellate practice (at the Board of Patent Appeals and Interferences and the Court of Appeals for the Federal Court). Her litigation practice includes the development of trial strategy and participation in discovery, with extensive involvement in preparation of expert witnesses in technologies involving genetic engineering, molecular biology, drug screening, and pharmaceuticals. Most recently, in an appeal she argued at the Court of Appeals for the Federal Circuit, Dr. Coruzzi and her team at Pennie & Edmonds won a reversal of a jury verdict and invalidated a patent covering cell-based assays for drug screening in Sibia Neurosciences Inc. v. Cadus Pharmaceutical Corporation, 225 F.3d 1349 (Fed. Cir. 2000).

Dr. Coruzzi was the chair of Pennie's biotechnology and pharmaceutical group, and oversaw the day-to-day management of the group by working with the core biotech partners. She also handled associate recruitment, training, work assignment, and marketing matters. Furthermore, she organizes biotech seminars that serve as a forum for disseminating case information, promoting discussion and debate, and generating strategic planning ideas.

Dr. Coruzzi is a frequent invited speaker at symposia on patent law issues related to cloning, genomics, ESTs (expressed sequence tags), SNPs (single nucleotide polymorphisms), and gene therapy. She is a member of the New York State Bar Association, the American Intellectual Property Law Association, the New York Intellectual Property Law Association, and the American Association for the Advancement of Science. She is also a member of the Metropolitan Opera Guild, the Metropolitan Museum of Art, the Museum of Modern Art, and the American Museum of Natural History.

Successful Biotechnology Companies: Developing the Legal Strategy

Matthew D. Powers
Head of the Global Patent Litigation Group
Weil, Gotshal & Manges LLP

There are two types of successful biotechnology companies: Those that succeed with the aid of a good legal strategy and those that succeed despite their legal strategy. To be part of the former group and develop a legal strategy that will contribute to a biotech company's success, several key ingredients are required. First, the company has to appreciate the importance of intellectual property rights to the competitive landscape in the biotech space. Second, the legal strategy must be customized to the company's technology assets with input from those who best know the technology. Third, the legal strategy must build on the company's relationships with key customers and enabling partners in technology development, marketing, and finance. Fourth, strategy, much like technology, should not stand still. Constant consideration and reconsideration of legal strategy is required. Finally, the strategy must fit the personality of the company, and not the other way around. Below, we describe a proven path for the development of a successful legal strategy for biotech companies.

Buy-In from Key Executives

A common problem faced by promising biotech companies is that they too often are reactive rather than proactive in the face of intellectual property issues. Having focused on technology for their entire careers, the scientists who head these companies often believe that good technology is enough to create a successful company. Even the biotech companies that do focus proactively on their own intellectual property often fail to invest sufficient effort in avoiding or overcoming the intellectual property roadblocks created by their competitors. In the end, of course, to benefit fully from successful investments in technology and intellectual property, a company must be known in the industry as a company willing to litigate whenever necessary.

We have found it important early in our relationship with a given company to meet with the key executives to ensure that they appreciate the importance of intellectual property. Normally, if the bright, entrepreneurial professionals who manage biotech companies do not already appreciate the importance of a proactive intellectual property strategy based on their own experiences, they can easily be educated about its value.

Of course, not everyone in the executive suite is going to have the same affinity or aptitude for intellectual property affairs. It is thus beneficial to identify at least one senior executive who has the skill set, as well as the desire, to be involved in intellectual property issues. Such a person should make for a good witness in the event that intellectual property litigation of one kind or another becomes necessary. That person can then be positioned as the corporate "spokesperson" on issues such as the reasonable basis for the company's actions. This can help insulate other executives who may be uncomfortable or ill-suited for litigation.

Choosing an influential management team member to remain close to and savvy about intellectual property in general (and litigation specifically) can also help the negotiation of the business terms of agreements entered into against the backdrop of complex intellectual property issues. This person need not, and often will not, be involved from day to day in managing the nuts and bolts of the legal or intellectual property function, but ultimately will prove a strong asset when relevant situations arise.

In identifying an executive suited for a high-level intellectual property role, it is important to avoid simply choosing the person the company is inclined to designate for that role. In companies where intellectual property is undervalued, a less effective or underutilized member of the management team is often designated as the intellectual property interface. In other cases, that role defaults to the person responsible for the legal function in general. However, the person responsible for the legal function is often not someone who is empowered to make key business calls on important deals that have intellectual property considerations at their heart. Likewise, if matters go to litigation, a legal representative of the company is not necessarily going to be viewed favorably by juries that are often skeptical of lawyers. Indeed, juries are often turned off by a company that cedes the ultimate business decisions about intellectual property, such as whether the company should stop selling products in view of a patent infringement assertion; this decision will naturally be considered by an attorney differently than it would be evaluated by a businessman. Further, an attorney-witness normally presents a host of thorny privilege issues that are best avoided. We have found that persuasive and credible representatives can be found in a host of corporate functions, including the chief executive's suite, and that the executive's aptitude and interest can be the most important factor in making the right selection.

Further, we have found that biotech companies that succeed because of their intellectual strategy have at least one senior executive who takes a special interest in and has a particular aptitude for intellectual property issues. The importance of such a key resource is placed in stark relief when the company is involved in high-stakes intellectual property litigation. Because most successful biotech companies are involved in such litigation, emerging biotech companies should ensure that they have an intellectual property-savvy executive in place as an important piece of any strategy for success.

Another value of having a senior executive supporting and involved in a proactive intellectual property strategy is that it increases the chances that the company will invest sufficient resources to fund an intelligent and long-term set of policies. Policies requiring a significant investment include, for example, effectively patenting research at an early enough stage, fully considering whether exclusivity is more desirable than licensing, and having the fortitude to pursue litigation when necessary.

In sum, a great intellectual property strategy will not be worth much if the necessary resources to implement and execute the strategy are not invested. In our experience, we have found that senior executive buy-in is a particularly reliable way to ensure that the proper resources are invested.

One Size Does Not Fit All

While it is natural to build upon experience, the same intellectual property strategy will not be a fit for two different biotech companies. Indeed, we have found that the same strategy is often inappropriate even for two parts of the same business. When we advise biotech companies, we make sure we learn enough about the company to customize our advice. To do this successfully, it is vital to understand not only the technology already developed by the company, but the future technology roadmap of the company. Often, we see intellectual property strategies that appear to be nothing more than the result of prior experience with historical products or technologies. The development of intellectual property strategy cannot trail the development of the technology it is supposed to protect.

Focused effort is required to understand a company's future technology path and the markets in which its future generations of products will compete. It is important not to accept as gospel the company's pre-packaged presentations that they are used to giving to investors and customers. Too often, legal advisors rely upon this, which is a mistake. Another common mistake is accepting information about the company's technology and market strategy from a single person who has been assigned to interface with the legal team; this person has been assigned to insulate the rest of the technical staff from legal matters, and he or she is not necessarily the most reliable resource upon which to develop a comprehensive legal strategy.

As legal consultants, we meet with and ask probing questions of several key technologists at the company and combine these answers with the benefit of our own technical expertise. Often, different employees of a company will have different views about the technology direction of this company and the likely areas of success. In our experience, each will have something to add. It is then incumbent on the legal team to evaluate how the technology path of the company should affect its strategy.

In litigation, an early understanding about the technology direction of the company is particularly important. With such information, it is possible to create the best arguments on the merits of the dispute. For example, by understanding the technology early, the legal team can begin to consider strategies to end the litigation earlier via a knockout punch, if one exists. For example, biotechnology companies often operate in a very complex licensing environment, which can support a license defense that will narrow or eliminate key claims. In addition, because many judges are willing to define the scope of patent rights early in the life of a litigation in a procedure commonly known in patent litigation circles as a "Markman hearing," there is frequently an opportunity to position the case for an early and attractive resolution. To do so, however, one must have a command of where the company's technology is at present, as well as where it is going. This understanding must be sufficient to explain to the court why an early decision will have the potential to resolve the matter effectively.

Knowledge about the future technology roadmap of the company can also create a strategy that will open up an end run around the litigation process. On several occasions, we have brainstormed with biotech clients to create new technologies that avoided intellectual property rights being asserted against the company. To our delight, those design-around products actually worked better than the existing technology that had been accused of infringement. This was accomplished by understanding what customers wanted of future products and what the company could do to improve the product within its technology base. Relative to many other technologies, biochemical molecules are often readily evolved in a way that can make them work better while also avoiding infringement.

Finally, successfully customizing a company's strategy requires a deep appreciation of the company's technology portfolio. In our experience, the required time and effort to gain such an appreciation are a solid long-term investment.

Relationships Matter

Knowledge of a company's technology is vital, but not sufficient, to develop successful legal strategy. A company's relationships with its customers, vendors, development and marketing partners, and others provide the context in which the company's strategy must be shaped. Much as a legal advisor needs to understand a company's technology, the company's relationships and the direction of those relationships must also be well understood.

For a company that develops its own technology, a key question is whether to attempt to retain exclusive control over the products that will embody that technology. From a litigation perspective, an intellectual property owner that retains exclusive control over its rights, all other things being equal, will be entitled to a greater recovery for infringement than a company that freely licenses its rights. Likewise, a party that retains exclusive control over its intellectual property will have more success securing injunctions against infringers' use of its technology. However, the network of relationships necessary to support an exclusivity strategy is very different from the network of relationships necessary to support a broad licensing strategy, which is something that must be carefully considered.

For a company that licenses the technology it uses for its products, rather than developing it internally, the specifics of its relationships with its key technology suppliers can be critical. Again, the option of exclusivity is a desirable goal from a litigation perspective. In addition, it is generally desirable to retain control over the intellectual property rights, such as control over the patent prosecution process. Different technology suppliers will treat these issues very differently.

Another relationship issue that arises frequently for biotech companies is how the patent claims are framed. The entity that controls the procurement of the patents also has substantial discretion in the form the patent rights will take. Virtually any technology can be covered by a patent as a method, an apparatus, a chemical composition, a kit, or a product by process. For example, a patent on a molecule can just as easily be sought and obtained as a patent on a method of using the molecule. Because of this flexibility, it is important to ensure that the claims are in the best form for the company's relationships. A method of use may be more appropriate for suing end users while a method of manufacture may be more appropriate for suing the companies that synthesize chemical compositions. Each biotech company should think ahead to whom it would want to be positioned to sue insofar as such action becomes necessary. Too often, we have evaluated situations where the mere form of the claim created major difficulties for the patent holder, because the form was directed to a party that the company was disinterested in suing or created formidable proof problems.

In every situation, the nexus between the company's relationships and its legal positions will vary. The key to effective legal strategy is to make sure the company's relationships are taken into account fully, even when ostensibly prosaic decisions, such as the form of the patent claims, are made.

Do Not Stand Still

In intellectual property strategy, as in life, it is critical to avoid standing still: instead, be constantly evolving. A successful strategy for a startup company is unlikely to be the same as that for an emerging growth company that has promising products in the regulatory pipeline. A company with successful products that have been through regulatory approvals will warrant a further evolution of strategy.

To avoid the scourge of inertia, we often find it useful for companies to schedule periodic legal strategy reviews that step back from the daily issues that confront dynamic companies. At such sessions, which are often timed with other "off-site" retreats, serious issues of long-term intellectual property strategy can be given due consideration. These sessions essentially counteract the tendency of managers who often operate on an "inbox" basis, prioritizing the items in the latest e-mail they receive. Blocking out time and creating expectations that management will focus on high-level legal strategy is a good way to avoid this phenomena.

Another tool that ensures adequate attention to the evolution of legal strategy is to provide companies with an audit of what they are doing on the intellectual property front, and what they should consider doing. Indeed, experience has proven that this is one of the highest value services we can provide. Intellectual property strategy is unlikely to evolve on its own and keep pace with the evolution of a dynamic biotech company. Accordingly, special efforts need to be made to keep pace in this critical area.

The Personality Must Fit

Years of experience have demonstrated that developing intellectual property strategy is an art, not a science. As mentioned, an intellectual property strategy that fits one company rarely fits a similar company with a different personality. We normally attempt to understand a company's personality before recommending particular strategies.

One key character trait is a company's appetite for risk. A company that is more aggressive is more likely to pursue strategies that focus on capturing exclusivity. Likewise, aggressive companies are, on average, more comfortable pursuing a strategy that may involve the threat, explicit or otherwise, of litigation. As an interesting aside, we have found that companies that are more self-confident about litigation and are more willing to accept the litigation path do not necessarily end up in more litigation. In fact, such an approach often deters litigation.

Another key character trait of a company is whether it is "open" or "closed." For example, open companies believe that proselytizing the community about the technology they work on is the best way to enhance the likelihood that it will be adopted early and widely. Closed companies are more likely to want to surprise competitors with their developments to obtain a head start. Public companies may tend to be more open, and private companies more closed. In any event, a closed company may be better suited for a trade secret strategy, rather than a patent strategy. Now that patent applications are generally published before patent issues arise, the patent process is likely to result in broad dissemination of a company's technology—with no guarantee that a patent will ever issue. Although trade secret protection can be easier to lose, it also allows a biotech company the opportunity to keep its development work confidential for a longer period of time. A closed company may also be more averse to litigation. Notwithstanding the presence of protective orders in most biotechnology cases, litigation has a way of forcing the disclosure of information a company may not want to disclose.

In sum, the process of developing a successful intellectual property strategy involves the consideration of a host of factors, many of which are quite fact-dependent. Keeping in mind the company's personality when shaping its intellectual property strategy is a particularly important component of strategy development.

Effective intellectual property strategy does not just happen. In the biotech field, an effective and winning strategy is generally more important than in other fields because of the density of patent and regulatory issues. Yet it is often harder to develop and keep current because of the complex and dynamic nature of biotechnology. This chapter outlined some of the tools and tactics we use to help biotech companies develop and implement a successful strategy. Following this advice should increase the likelihood that a biotech company's intellectual property strategy advances its success, rather than hinders it.

Matthew D. Powers is the head of Weil, Gotshal & Manges LLP's Global Patent Litigation Group and specializes in trying patent and trade secret cases. He has litigated and tried cases all over the world involving a wide range of technologies, including semiconductor devices and manufacturing equipment and processes, DNA sequencing, medical devices and other biotechnologies, computer hardware and software, communications (network and telephony), and the Internet.

Mr. Powers is an editor in chief of the Intellectual Property & Technology Law Journal, *and has published extensively on various aspects of intellectual property law and litigation. He is a frequent lecturer nationally and internationally on intellectual property litigation issues. Mr. Powers also teaches a patent litigation course at the University of California, Berkeley's Boalt Hall School of Law, and has lectured on patent law at Stanford University and Santa Clara University. He serves on the firm's overall management committee.*

Mr. Powers received his J.D. from Harvard Law School and a B.S. from Northwestern University.

Counseling Biotech Companies in a Competitive Market: The Interplay Between Freedom to Operate and Patent Protection

Eric S. Furman, Ph.D., J.D.
Partner
Knobbe Martens Olson & Bear LLP

Getting Started

When initially sitting down with a client, I ask them to describe the landscape in which they operate. Who are the big players, the 800-pound gorillas in the field—their competitors? I explain that understanding the competition, where they are going, and following the competitor's intellectual property portfolio are critical to developing a successful research and development program of their own, and a successful intellectual property position in the field. After finding out who the competitors are—and their strength (e.g., the willingness to litigate, depth of competitors' intellectual property portfolio)—I ask the client how they intend to enter the market, what their niche is, and what intellectual property they intend to carve out of the landscape of others' rights. In addition, I ask whether they want to sell a particular product, or do they instead intend to be an intellectual property clearinghouse (i.e., a licensor of technology), and how do they intend to get from where they are now to that point. It is surprising to me how many companies enter into vastly expensive research and development programs without looking first to determine whether the end product/result is commercially feasible. That is, "freedom to operate" in the field of interest and the number of licenses, opinion letters, and the like, should be carefully considered before embarking on an expensive research and development program, because the company may not be able to sell its product at a rate the public/industry will bear.

For example, hypothetically a client comes to you with a diagnostic test for a congenital disease. They have developed an amplification-based screening technique that allows a user to rapidly identify a number of mutations in a gene that has been associated with the disease. A quick review of the patent literature reveals that someone owns the rights to the enzyme used for the amplification technique, someone owns the right to the nucleic acid encoding the gene, someone owns the right to the mutations or methods of screening for the mutations to diagnose the disease, and someone owns the rights to the automated screening technique used to obtain the result. The bottom line is that your client will need to obtain at least four licenses to be able to practice the diagnostic test. These licenses may make the costs of performing the assay not competitive with others in the field. Also, one or more of the four players may not want to license their technology, especially if they are a competing diagnostic company. Such an experience could break a startup company; however, had they chosen a different gene or a slightly different technique, they may have been able to carve out a niche that would get them off the ground with an affordable diagnostic fest.

Too often in the biotech field, companies begin operating on a research plan alone without careful thought of how to achieve the end result, be it a product on the market or a licensing position that offers the least resistance. Since a lot, if not the majority, of biotech startups stem from scientists of academia, it should be realized that academic purpose and commercial purpose do not always coincide. Typically, people take an approach they are familiar with, but sometimes it is necessary to retrain the focus from an academic-style approach to an approach focused on a commercially viable enterprise. Although everyone desires to have a profitable business, the approach often seen is one academicians are most familiar with—grant proposals, material transfer agreements, collaborations with other scientists at educational institutes, and the like—rather than a concrete business plan that takes into account freedom to operate and the difficulties involved in coordinating multiple licenses at a rate that is economically feasible; academicians also can overlook the ultimate threat, an inability to obtain a license to an aspect of technology that is essential to the development of the lead compound. Research and development is essential, but the focus needs to be on a commercially viable product and the costs associated with obtaining that result. The race is for protection of a profitable technology, one that is free of the rights of others and for which a strong patent position can be maintained.

It should also be made abundantly clear to the client early on that a "freedom to operate" analysis is not an assurance or a guarantee that the client will have the unfettered right to make, use, or sell their products. The client must realize that a freedom to operate analysis only reduces the risk of learning in the future that the client is unable to practice the invention without interfering with the rights of a third party. Too often, clients and investors are looking for a guarantee that they will not have to take a license, or that they will not get sued for patent infringement. Because of the nature of the investigation, the tools that are used, and the budgets set by the client that necessarily limit the scope of the investigation, such assurances cannot be given. An investigation could range from an analysis of specific patents identified by the client to searches for blocking patents in public and private databases. Often, these databases are replete with errors and omissions. In addition, the proper scope of the claims of any particular patent cannot be fully appreciated until the file history is ordered and reviewed. This process can be extremely costly, and clients should balance their aversion to risk with their budget. Although some clients can afford a freedom to operate analysis that is conducted in every country in which the product will be sold, most clients have a limited budget and such expenditure is not realistic. Many misunderstandings can be circumvented by clearly identifying to the client the scope of the analysis that will be performed, the limitations that are inherent in the process, and the amount of reliance a client can place on the findings.

Structuring the Client Relationship in a Competitive Field

The client needs to understand up front that information on the field and their competitors is necessary. In general terms, I encourage clients to carefully monitor discoveries in the field and the activities of their competitors. I encourage the client to attend tradeshows and conferences where their competitors discuss their research, but I have found monitoring the publication of patent applications to be a much better approach. Sophisticated companies screen their talks, presentations, disclosures, and the like to carefully ensure that no unprotected intellectual property leaks out to the public. As a consequence, what you hear at a meeting is generally very old news. Since patent applications are published eighteen months from their earliest priority date, the cutting edge of the field can be better monitored by following these publications. Commercial services are now available that allow one to monitor the publication of patent applications that list specific assignees, keywords, patent numbers, inventors, and the like. The cost of setting up and maintaining these types of monitoring services is relatively inexpensive and can be invaluable for developing a successful patent strategy. A little time spent searching patent and literature databases to identify past publications and issued patents of key inventors, assignees, and the subject matter of the field is also money well spent. Lastly, I encourage companies that deal with genes and proteins to perform a sophisticated BLAST analysis on their sequences to identify common fragments, domains, and competitors that are in the field but unknown to the client. The BLAST analysis also allows the practitioner to craft better claims, which saves money in amendments and lost time. Knowledge of where your intellectual property stacks up against the prior art and the competition is essential.

For example, hypothetically, a client has developed a fusion protein that utilizes a linker that is a fragment of a particular gene thought to be in the public domain due to a public disclosure. A BLAST analysis reveals not only the full-length gene from which the linker was derived, but several other genes that contain either the full-length linker sequence or fragments thereof. A search of these covered sequences and the names of individuals associated with the submission reveals several United States patents that reference the sequence. One of the subject patents is found that predate the believed public disclosure, and the patent contains claims to fragments of a gene that contain the linker sequence. A freedom to operate issue has arisen and an opinion letter as to the validity of the claims or a different design is advisable.

In discussing patent strategy with a client, it is important to provide all of the options that are available under a set of circumstances and provide them with enough information to allow them to make the decision. Although it is important to provide them with a recommended path, it is more important to educate them as to why that path is the best alternative. At the first action received by a patent office—an international search report, for example—the art cited against the patentability of the claims should be carefully analyzed and reported to the client so the arguments against the novelty and obviousness of the invention can be immediately considered by the client. It may be that a dead end to a particular project has been reached, or that a considerable amount of money will have to be spent to obtain a patent that may be easily designed around. Because these issues can have a dramatic effect on a business, they should be discussed and analyzed early on in the patent process.

Difficulties Faced by the Biotech Industry

The hot topics in biotech patent law today are written description, enablement, and restriction practice. There are always interesting aspects of biotech patent law concerning the loss of patent rights due to prior art issues, in particular, the subtleties of inherency law and approaches to overcome or insulate your invention from such problems. However, a series of biotech cases over the last few years have been narrowly interpreted by the Patent Office, and many patent examiners are forcing applicants to limit the scope of their claims by arguing that the applicant did not adequately describe the full scope of the claimed invention, an aspect of 35 U.S.C. Section 112, the "written description" requirement,[1] or that one of the skills in the art would have to perform too many experiments to possess the full scope of the claimed invention, another aspect of 35 U.S.C. Section 112, the "enablement" requirement.[2] The *coup de grace* for many biotech patent applicants, however, is a quasi-combination of the written description and enablement requirement, affectionately referred to as the "super-enablement" standard.[3] That is, with respect to some biotech inventions, examiners argue that 35 U.S.C. Section 112 also requires that applicants demonstrate that they are in actual possession of the invention at the time of filing.

For example, hypothetically, an applicant has described in their application that fragments of a peptide, which are at least three, five, seven, ten, or twenty consecutive amino acids of the peptide, effectively inhibit cell differentiation. The application also provides data showing particular fragments that are three, five, seven, and ten consecutive amino acids in length, all having the XYZ amino acid motif, inhibit cell differentiation. Further, the applicants provide data showing that pools of chemically synthesized fragments of the peptide lacking the XYZ motif (twenty amino acids in length and five fragments/pools), inhibit cell differentiation. Although claims to the specific fragments comprising the XYZ motif are allowable, claims to a fragment of the peptide that is at least twenty amino acids in length but lacking the XYZ motif are not. The examiner argues that although it is evident that the applicant is in possession of a pool of fragments that inhibit cell differentiation, the applicant is not in possession of a fragment of the peptide at least twenty amino acids in length that inhibits cell differentiation, because the applicant did not identify which one or more of the five members of the pool are effective. Despite fully describing each of the peptides in the pool by sequence and providing

---

1 *See Regents of the University of California v. Eli Lilly and Co.*, 119 F.3d 1559 (Fed. Cir. 1997) and *Enzo Biochem, Inc. v. Gen-Probe, Inc.*, 296 F.3d 1316 (Fed. Cir. 2002).

2 *See In re Wands*, 858 F.2d 731 (Fed. Cir. 1988) and *Amgen, Inc. v. Chugai Pharm. Co.*, 927 F.2d 1200 (Fed. Cir. 1991).

3 *See University of Rochester v GD Searle & Co.*, 249 F. Supp. 2d 216 (WDNY 2003).

evidence that at least one member of the pool is effective, many biotech examiners are steadfast on the requirement that the applicant must specifically identify the embodiments that work from those that do not.

Also tied with these often insurmountable obstacles under 35 U.S.C. Section 112 is the progressively narrowing restriction requirement, a tool by which examiners reduce the complexity of their case load and narrow the scope of protection available to the biotech industry in the process. To combat this problem, biotech patent attorneys should carefully construct claim sets to take advantage of linking claim practice, and make the examiner aware that the pending claim set is subject to linking claim practice.

For example, hypothetically, an applicant claims a composition comprising a novel adjuvant and a cancer antigen. In claims that depend on the aforementioned claim, the applicant recites that the cancer antigen can be any one of a number of antigens from various viruses, and ten different sequences are identified by sequence identifier number. In the restriction requirement, the examiner requires that the applicant elect a single sequence for prosecution, and notes that the election is not a species election but rather that each composition comprising the novel adjuvant and sequence is an independent invention. The trap is set, and if the biotech patent attorney is not careful, the opportunity to have the broad claim reciting the composition comprising a novel adjuvant and a cancer antigen will not be examined. The biotech patent attorney needs to point out that the claims are subject to linking claim practice, the election is made for examination purposes only, and upon a finding that the elected composition comprising the novel adjuvant and sequence is found allowable, the examiner will consider examination of the broad claim reciting a composition comprising a novel adjuvant and a cancer antigen.

Another important issue for biotechnology today concerns the difficulty in raising money. There are several ways to help a client in this regard, including:

1. Organize the client's intellectual property portfolio so that a clear picture of the strategy and market path can be realized by investors.

2. Present not only rejections and approvals from the patent offices, but also proposed arguments to counter the rejections so a clear understanding of the patent position can be discerned. Realize that much of this information will be transferred to investors.
3. Present freedom to operate issues, competitors, competitors' intellectual property positions, and the need for licenses and/or opinion work to clear the path to market early so investors can better appreciate the environment.
4. Help the client develop a good marketing plan. A clear picture as to how the company is going to generate money is of primary importance to investors. For example, is the company selling product or licensing intellectual property?
5. Present realistic goals given the marketplace and the capital a company has to work with. The goals should mesh the current intellectual property strategy and the freedom to operate concerns.

When dealing with mergers and acquisitions (M&A), other issues should also be considered:

1. Analyze the available intellectual property carefully. Does it provide the protection for the acquiring company that is needed, or does it just talk about it?
2. Are continuation or divisional applications available so the acquiring company can craft claims to better protect its interest? Flexibility in the future is essential. Is other protection needed to secure the acquiring company's position in the field (e.g., licenses)?
3. Investigate all applications and publications by key inventors for the acquired technology. Are the inventors of the acquired company keeping the best technology out of the deal (e.g., in a spin-off company)? Have the inventors destroyed or jeopardized rights by early publication or offers for sale?
4. Check chain of title on applications and publications. Are there inventorship issues that cloud title?
5. Check the licenses of the acquired company to determine if they can convey rights (e.g., sublicense) to the acquiring company and the costs to practice the technology. How many licenses are required to get to market?

6. Check competitors' technology, patents, and patent applications and any opinion letters that are available to determine if more licenses are needed to get to market, the cost of operation in the field, and the exposure to liability.

Protecting the Intellectual Property

In designing an approach to developing a patent portfolio, I subscribe to the belief that more is better. A web of overlapping patent rights is a much stronger position than having a few key patents to protect the core technology. I advise filing multiple provisional patent applications and continuations-in-part while the technology is being developed. Early, during development of the technology, one should file multiple provisional patent applications. At each step in the process, priority dates are being established, and at the one-year anniversary of the earliest priority date, an international application that designates all countries, including the U.S., is filed and priority is claimed to all of the provisional applications. By protecting the invention at all stages of development, one not only acquires incremental provisional protection for the technology, but one instills value in the patent that issues because it is more difficult to invalidate a patent with multiple priority dates than one with a single date. Not only is it more costly to obtain and analyze multiple file histories, but many more arguments for date of invention must be analyzed. Similarly, by filing continuation-in-part applications with extensive priority claims, one makes a stronger patent portfolio by making the patents more difficult to invalidate. A web of priority claims and interrelated patent rights makes it much more difficult for a competitor to realize the scope of protected technology, and this uncertainty drives away the competition. Additionally, by maintaining a pending continuing application, the client is able to draft claims specific to competitors and maintain flexibility. These are factors that are lucrative to a potential licensee.

Preparing and filing extensive information disclosure statements that list all of the prior art found during the monitoring of the competitors is also necessary, as is cross-referencing information disclosure statements in related applications. Not only is it required by law to provide references that are material to the patentability of the claimed invention, but this process builds a stronger patent because an issued patent is presumed valid over the references cited. Furthermore, extensive information disclosure statements raise the cost of obtaining the file history for the application, which may deter some competitors from investigating the patent position altogether (e.g., the file histories for some patents and related documents can run several thousands of dollars just to obtain a copy).

Monitoring competitors' activities through the patent publication databases and the Internet is essential, as discussed *supra.* Several Web sites for plant field trials or animal clinical trials, for example, are available, and these links describe in detail the technology being tested. By staying advised of developments in the field, a company can focus its attention on specific competitors, analyze the competitor's portfolio for weaknesses, and exploit them. For example, one can identify key technology of a competitor and obtain blocking rights (e.g., patents that block expansion in the field either through license or assignment). One can then contact the competitor and force a cross-license or stop the competitor from expansion in the area. Additionally, by monitoring publications of applications, a client can insulate themselves from a competitor by monitoring prosecution of the competitor's applications and conducting prior art searches against them. Once key prior art to the competitors' applications are found, it can be sent to the examiners of the particular applications and by registered mail to the competitor. This forces the competitor to submit the references (to prevent a charge of inequitable conduct, which would invalidate the patent). It adds significant cost to their prosecution and slows things down considerably. One can also send competitors a copy of your client's issued patents, which also slows their progress and forces them to spend more money in analyzing your intellectual property, rather than creating their own.

In developing an international strategy, I recommend filing multiple provisional patent applications as the technology is being developed, as described before, and filing an international application that claims priority to all of the filed applications. At eighteen months from the filing of the first provisional patent application, the client must decide which countries designated in the international application they intend to file national applications in. It is important, at this point, to carefully consider the nature of the client's business and the countries that are the source of significant competition. It is also important to consider the legal regimes that are in place in the particular countries of interest. That is, it may be too costly in time or money to litigate a case in some countries; therefore, one should not file in that country.

The biggest issue faced is money. It is a difficult predicament, because it requires money to fund the research and development and patent protection. They go hand in hand in that the more patent protection one obtains, the more confidence investors have in the company; however, it is often difficult to obtain enough seed capital to get the research and development and patent programs going. Difficult choices in product development and protection must be made, and not all candidate projects are going to make the cut.

Technology Licensing

Again, the key point here is that one needs to be informed of the marketplace. One needs to know what the competitors are doing. By staying informed of progress in the field, one can carefully select appropriate targets for license or collaborations that may lead to a license. Once a web of patent protection is established, for example, informative letters can be sent to companies that appear to be using your client's technology. These letters inform the alleged infringer of the client's intellectual property rights and provide the infringer with an invitation to a license. Depending on the client's market position, they can also approach infringers with an invitation to collaborate using the client's technology to validate the company's candidates, or make the company's products more effective, or to provide the company with an alternative use for their technology. By initiating contact with competitors, oftentimes, research agreements can be forged, which ultimately lead to a license.

Too often, companies take the approach that this is a zero-sum game, where there is only one winner. This is not the case. Strength is obtained by forming alliances between companies, especially competitors, so that money can be spent on research and development rather than patent disputes.

Eric S. Furman, Ph.D., J.D. is a partner in the San Diego office of Knobbe, Martens Olson & Bear, LLP. He specializes in patent protection and other forms of intellectual property protection for biotechnology-related inventions.

Dr. Furman's practice includes strategic patent counseling and procurement, freedom to operate analysis, counseling on infringement and invalidity issues, licensing, and due diligence studies as related to negotiations for mergers and acquisitions. Dr. Furman currently represents clients with technology in the fields of genetically modified plants, antiviral therapeutics, dietary supplements, and diagnostics.

Prior to joining the firm, Dr. Furman was a research assistant in the laboratory of Dr. Leroy Hood, where he made transgenic mice with aberrant forms of myelin basic protein. As a doctoral student at the University of California, Los Angeles Medical School, department of biological chemistry, he characterized transcription and splicing complexes. While attending law school at the University of Washington, Dr. Furman was a teaching assistant for Donald S. Chisum, the author of the twenty-volume reference text on patent law, Chisum on Patents.

Practical Advice for the Biotech Entrepreneur

David Hoffmeister

Partner

Wilson Sonsini Goodrich & Rosati, P.C.

Having All of the Pieces in Place

The hurdles and challenges involved in building and sustaining a biotech company are tremendous. At the outset, there are three essential areas in which entrepreneurs must focus: developing and implementing a sound business strategy, assembling an experienced management team to execute the business plan, and managing the risks associated with operating a company in a highly complex regulatory environment. The road is not an easy one to navigate, but this is where entrepreneurs must begin the journey.

The Need for a Solid Business Plan

Bringing a new biotechnology or pharmaceutical product to market is an extremely expensive proposition, involving tens, if not hundreds, of millions of dollars. If an entrepreneur has any hope of raising financing from the venture capital community, the first step is to develop and implement a comprehensive and well-designed business plan.

In any given month, a venture firm will receive and review hundreds of business plans. Of these, only about five or so are seriously considered, with term sheets extended to only one or two companies. Therefore, a strong business plan—one that thoroughly analyzes all aspects of the business—is essential. A well-written plan will describe the product rational, how it will fit in clinical practice, the competitors, and reimbursement issues such as who will pay for the drug, the likely reimbursement rate, and challenges associated with reimbursement. Another component is realistic market projections and well-articulated, reasonable expectations on the timing for regulatory approval. Finally, the plan must also adequately cover the product's patent protection, licensing status, and any marketing exclusivity that will be granted by regulatory authorities if the drug makes it to market.

Legal Diligence and Requisite Preparation

Prior to making an investment, a venture firm will diligently investigate the material claims a company makes. Venture capital firms routinely turn to legal counsel, who are regulatory, patent, and licensing experts, to perform diligence on the technology and thoroughly evaluate the patent and licensing portfolio, reimbursement strategy, regulatory pathway to market, product manufacturing, and clinical details, as well as competition aspects of the business. Additional experts, including physicians and medical thought leaders, will be contacted for their opinions on the product and its clinical potential.

Consequently, a young biotech company should anticipate and prepare for this thorough evaluation by also hiring experienced counsel to run the transaction, manage the diligence, and negotiate the finer points of the deal. In fact, counsel should be hired well before the offering to allow adequate time to be educated on the overall product development strategy and the needs and expectations of management. At this point, areas of the business that need to be developed, as well as weak points where the company lacks expertise, should be identified, and the gaps should be filled with input from consultants and business partners. It is also valuable to generate well-developed, scientifically supportable responses to questions that are certain to arise during a presentation to a venture capital audience.

For example:

- What are the rate limiting factors that must be addressed before the product can be generated in commercial volumes by existing manufacturing facilities?
- What are the weaknesses of the company's patent position, and what are the opinions of patent counsel?
- What is the reimbursement landscape for the product, including the status with Medicare and private payors?

All of these questions are likely to arise. Technically rich, defensible, and accurate responses—supported by experts in the field—should be developed. Diligence of this variety, but more thorough, will occur should a company attempt to raise funds in the public market. Consequently, going through this preparation is always a useful exercise.

Management

Even with a solid business strategy, a novel technology platform, and defensible positions, a company will not go far without strong leadership. An experienced management team is particularly important for a young company to attract both early and late rounds of financing. An experienced chief executive officer who has been in the biotechnology or pharmaceutical industry with a track record for bringing products to market is important. Entrepreneurs in these positions often come from large pharmaceutical companies having served as division presidents, vice presidents of sales or marketing, or having some well-founded knowledge of the industry. Moreover, young biotech companies will be required to tap public markets for funding fairly early in the product development process, because large scale clinical trials are expensive, and the venture capital community is willing to invest only so much money. After securing between $30 and $50 million in venture financing, companies will need to consider the public equity markets or actively seek a development partner. To be as marketable as possible, an experienced management team, including a top-notch chief financial officer, regulatory and manufacturing vice presidents, and general counsel, must be in place. This management team must be able to fully manage the strengths and weaknesses of the patent, regulatory, and clinical aspects of the development program, and effectively communicate and instill confidence with the investment community.

Regulatory and Legal Issues Specific to the Industry

The Regulatory Process

The complex regulatory regimen that governs the approval process creates legal issues unique to the biotechnology and pharmaceutical industry. For example, a patented compound is entitled to twenty years of patent life from the time of discovery. However, on average, it takes a new chemical entity about seventeen years of development, clinical study, and regulatory review before it can come to market. Clearly, the timeframe for re-cooping the research and development investment is limited, so fully understanding the regulatory landscape, including when generic competition is likely, and executing a well-designed and highly focused clinical development program from the time the company is formed is essential.

The process of securing regulatory approval is expensive, time-consuming, and complex. In reviewing these products to determine whether or not they should be approved and made available to the medical profession and patients, the FDA is accountable to numerous government and private organizations, as well as to scientific and medical communities. Most of the agency's principle decision-makers are physicians or scientists, to whom the respect of their professional peers is of great importance. When the FDA makes decisions about cutting-edge technologies, they must be respected as sensible in the relevant scientific and medical disciplines. To do so, the agency formally seeks the advice of its advisory committees, which are composed principally of leading academic physicians and scientists, and generally follows their advice. Although a large, critical part of its work is the review of new and potentially blockbuster products, the FDA is largely insulated from effective outside oversight. Its reviews are based on trade secret data that is generally not disclosed elsewhere (except to similar regulatory bodies in other countries), is highly technical, requires specialized competence in certain disciplines, and is extremely labor-intensive.

Because the FDA's reviews of marketing applications are critical to the fate of biotech companies, these companies have been traditionally reluctant to challenge any particular FDA product review decision in another forum. Resorting to the courts to challenge an FDA denial of approval of a new product is particularly unappealing for several reasons, including:

- The technical nature of the subject matter;
- Prevailing doctrines of administrative law, which routinely lead to a deference to administrative agencies on scientific and technical matters;
- The institutional limits on the judiciary, which is not well equipped to make judgments about scientific and technical matters; and
- The unattractiveness of trying to market a product that FDA has initially refused to approve.

The result is a limited but important area in which the FDA's exercise of power is, to a considerable extent, unchecked and unaccountable.

And Why it Matters

Young biotech companies and entrepreneurs need to understand this important dynamic and the significant regulatory risks it presents. The FDA is held accountable on those few occasions when a segment of the public takes a particularly strong interest in a pending review of a proposed new product, such as a new AIDS medication, or when a past approval has turned out to be a serious mistake, as in the withdrawal of the arthritis drug, Vioxx. In such circumstances, extraordinary public attention, including congressional hearings, is focused on the FDA's decision-making process, and the agency is called to publicly account for its decision. But in reality, the consuming public does not generally hear when a young company's product is not allowed onto the market, or when the FDA is requiring additional data that will require an investment of additional millions by the young company.

The Importance of Compliance

No industry in the world is regulated more than the biotechnology and pharmaceutical industry. Young companies developing novel products will be required to comply with a myriad of regulations, both before and after approval. A compliance infrastructure must be developed in the early stages of the company's existence and continue to be supported as a top priority throughout its life. Failure to take compliance seriously throughout all phases of the company's existence can be costly and, at times, threaten the viability of its operations.

It is also critical for entrepreneurs who may have come up through the corporate marketing or sales ranks to understand that the FDA is a law enforcement organization. It enforces the law through effective modes of publicity, a variety of administrative actions, and with the approval and assistance of the Justice Department, through judicial actions. The FDA takes enforcement actions to remove violative products from the market or prevent their distribution in the first place, to obtain compliance with FDA requirements, to punish persons responsible for causing violations, and to deter others from violating regulations in the same or similar ways. It has broad discretionary power to decide the violations against which it will take action; the form and depth of that action; and in doing so, which products, companies, or individuals to hold accountable.

For the most part, enforcement actions are built on evidence of noncompliance with FDA regulations that is gathered by investigators during inspections, and supplemented by information obtained from other sources, including competitors and whistle-blowers or disgruntled employees. Additionally, the agency may rely on information that a company submits to meet FDA regulations, like adverse reaction reports, or that a company submits voluntarily, like product recall information.

Over the past several years, the FDA has significantly altered its approach to enforcement actions. Congress has given the agency greater enforcement authority, and the agency has developed compliance policies and procedures that are more varied, more resource-efficient, and swifter. Consequently, the FDA's current enforcement policy is best described by the agency's greater use of administrative (rather than judicial) enforcement actions; greater coordination within the agency; greater coordination with state enforcement authorities, and professional and trade associations; greater use of adverse publicity; and the use of informal policy guidance documents to encourage companies to operate within the FDA's expectations. In the area of judicial enforcement, the FDA focuses on cases involving fraud, intentional or knowing violations of the laws, and cases involving product defects with significant consequences for the medical profession and consumers.

And Why it Matters

It is a myth that the FDA brings enforcement actions against only large multinational pharmaceutical companies. While it is true that enforcement actions against large companies tend to receive significant media publicity, young companies are by no means exempt. In fact, there are more enforcement actions initiated against small companies. A young company's failure to invest in compliance and take seriously its regulatory responsibilities would be shortsighted and unwise.

New Legal Developments

Looking ahead, change will be constant in the areas of generic products with an emphasis on allowing generic competition into the marketplace faster, which in turn will result in lower costs for medication to the American public. It is also just a matter of time before there is a regulatory path for generic biologics, which will narrow the window to recover research and development investments for branded biologic products, similar to that experienced by pharmaceuticals when subject to generic competition in 1984. Additionally, the export of cheaper drugs and biologics from Canada has been subject to much debate recently, and it is likely that new legislation will be passed allowing exports from Canada. These are important developments that young biotechnology companies must understand and take into consideration when establishing early strategy and managing the company as it matures.

Lastly, the broad anti-kickback laws enforced by the Office of Inspector General have resulted in fines collected from drug companies in the past couple of years that are approaching $2 billion. These fines have been associated with illegal marketing schemes and failure to accurately report prices to the government. Aggressive enforcement of these broad laws will continue, so young companies entering this industry must continuously evaluate compliance programs and educate their sales and marketing staff of the seriousness of violations. A conviction in this area will cost a company millions of dollars, and it may result in its product being excluded from the federal Medicare program, a dual-sided death sentence for many companies.

A Word of Advice

From the outset, entrepreneurs should surround themselves with people who have gone through the regulatory approval process, have successfully brought products to market, and who fully understand and appreciate the dynamics of this industry. They should seek advice from attorneys, accountants, bankers, and consultants who have documented life sciences expertise, and from individuals who have built and run successful biotechnology companies. Financial risks are too great in this industry not to seek advice from individuals who are extremely well versed in the regulatory landscape and the unique aspects of running a biotechnology company.

David Hoffmeister is a partner at the law firm of Wilson Sonsini Goodrich & Rosati, where he leads the firm's drug and device regulatory and health care law practice within the life sciences practice.

Mr. Hoffmeister brings more than eighteen years of experience in drug and device regulatory and health care law to the firm. He represents pharmaceutical, biotechnology, dietary supplement, medical device, and diagnostic clients and advises them on a variety of different health care and regulatory issues, including inspections, recalls, labeling, advertising and promotion, and strategies for obtaining FDA approval and clearance. Previously, Mr. Hoffmeister was senior counsel for drug and device law at Syntex Corporation, where his primary focus was to advise senior management on all worldwide issues affecting the corporation's and its affiliates' ability to develop, manufacture, and distribute pharmaceutical, device, diagnostics, and over-the-counter products in compliance with the Federal Food, Drug, and Cosmetics Ac, and implementing regulations, and other state and federal health care laws.

Mr. Hoffmeister received his J.D. degree from the San Francisco Law School and his B.S. degree from the University of the Pacific. He is admitted to practice in the state of California and before various federal courts, including the United States Supreme Court.

The Key Asset: How to Build a Great Board and Effective Board-Management Communication

Bruce W. Jenett
Co-Chair, Life Sciences Practice
Heller Ehrman LLP

From its very inception, the creation of a powerful and productive board should be viewed by management and the board itself as a critical and strategic company activity, one that represents a powerful way to build competitive advantage. Proactive thinking by both existing board members and the CEO about, and careful selection of, board candidates having the right background, experience, and personality to work as a team together and with management, can often make the difference between creating a successful company able to handle both problems and opportunities with flexibility and creativity, versus a company struggling to move ahead, burdened by bickering or apathy on the board, and tension between the board and management, which ultimately can (and often does) destroy the company.

In terms of the board's size, small boards (five to seven members maximum) are best; an odd numbered board is also desirable in the unusual event of a board deadlock.

When reviewing potential board members from investor groups, both existing board members and the CEO should focus on the quality, value, and personality of the individual who will serve from the relevant investor group, not just on the quality or value to the company of the investor group itself.

Both industry diversity and personality diversity on the board, within reason, is desirable. Ideally, each non-management board member should have a specific "virtual officer" expertise (such as technical, finance, regulatory, scientific development, marketing, or the like) upon which management should be able to draw. Further, it is useful to have the given director's "virtual officer" expertise be at a level that is two to three years ahead of that of the relevant management member of the company. This helps pull the company forward, and it allows management to anticipate and better handle opportunities and problems that may arise. For example, if the company is a couple of years away from an IPO, it may be useful to have as a board member a CEO of another company that has gone public fairly recently. This will help guide management in planning for the IPO and for operation as a public company.

It is also useful to have at least one board member who is a user of the company's technology. For example, if the company is making medical/surgical products, it can be helpful to have on the board someone reasonably high-ranking within an HMO who will be able to guide management in designing and marketing products that will be accepted and purchased by hospitals or other distribution outlets.

In further considering the constitution of the board, it is desirable to have a member from directly within the company's industry and, if possible, from a potential corporate partner (perhaps a retired executive of the desired partner company). It also may be useful to have at least one board member who is *not* from the industry in which the company is operating. For example, a software industry CEO might provide fresh insight on a medical products company board.

Finally, a "pretty face" (i.e., someone with "star power") is not enough to bring value to a board. Each board member, even well-known ones, must be willing to make the time and emotional commitment that will allow his or her particular skill set or level of experience to be useful to the company, both before and after its IPO.

The CEO/president, while usually on the board in a privately held company, should not be chairman of the board. Instead, the non-management chairman should play the role of "partner/coach" to the CEO/president. There are mixed views about the advisability of having the president or CEO on the board of a public company.

Different Stages, Different Boards

Pre-IPO boards usually are dominated, both in numbers and in personality, by venture capital investors, along with one or two outsiders, typically picked by the venture capitalist investors with CEO acceptance. As the company moves towards an IPO, boards tend to get restructured in order to add individuals who hold appeal for the public markets, not necessarily just to the venture capital community. This is particularly true, as noted below, in the current securities regulatory environment created under Sarbanes-Oxley. Boards in private companies are also typically restructured at major financings to make room for, or to add, individuals designated by the new investors. This change can sometimes radically alter the character and interaction of the board, both among each other and with the CEO and management. Sometimes that change is good, and sometimes it is not; regardless, management rarely has any input on this issue, as it is decided among the old and new investors.

Just as CEOs often get replaced with someone better experienced for the next stage of the company's growth and development (particularly as the company gets closer to an IPO), boards may also need to change in composition, and perhaps even in overall personality, to accommodate and be more attuned to the various major stages of growth of the company. The post-Series A Preferred financing board is rarely the right board for the company after a Series D Preferred financing or when the company is grooming itself to go public. Furthermore, given the impact and requirements of Sarbanes-Oxley, more and more private companies are looking well before the IPO for the relevant expertise and experience, particularly financial, required under Sarbanes-Oxley for a public company board. Many such companies have found that it is very difficult to recruit and bring onto the board new members with the right Sarbanes-Oxley qualifications in a timely manner before the IPO. Board members should look well ahead on these stage-related board composition issues and plan what is best for the company, not just for their own self-interest.

When a director is "tired" (i.e., passive or not really participating), the chairman, rather than the CEO/president, should speak with this director, and other key board members, to collectively determine whether it is time for that director to resign and be replaced.

Keys to Effective Board Functioning

The most important element in ensuring smooth and effective board activity is the creation of a board that understands that it must operate as a team, not as individuals, not only in terms of how they relate to one another as board members, but also, in a more fundamental sense, in terms of how the board interacts with management, primarily with the CEO.

Early on in the process of creating a board, even at the beginning of the company when the board typically is composed of founders and "friends of the family," certain board guidelines and rules should be set forth, with a promise from the board that it will respect and adhere to these policies. First and foremost, the minimum time commitment, both for board/committee meetings and for non-meeting involvement, should be clearly understood and agreed to by each board member.

Although desirable skill sets in board members will vary depending on the type of company and the constitution of the board itself, there are certain qualities in a board member that are never helpful. For example, passive or somnolent board members do not add value, and neither do cantankerous or obstreperous board members; subverting the CEO is not a useful role for a board member.

On the other hand, intellectually challenging (but not argumentative) board members are strong additions, as are collegial, supportive, decisive, and active directors. In order to promote this type of positive board behavior, management must be willing to be "coached" by the board when necessary; the CEO must set the example for other senior management to be open to and accept appropriate coaching from board members. However, in doing so, a board member should never usurp or weaken the managerial power and authority of the CEO vis-à-vis management, nor psychologically overwhelm line management with the board member's personality or expertise.

The Role of the Chairman of the Board

The chairman should provide periodic, meaningful, honest, but collegial feedback to the CEO from the board. This should take place outside of board meetings to prevent the CEO from being blindsided in a board meeting, or within the company, or in the marketplace. Further, this will help the CEO continue to grow in the position with respect to company strategy and tactics, "emotional" or "political" issues within the company, and issues involving management's dealings with the board.

Director Participation and Evaluation

Being an effective director requires appropriate preparation for, attendance at, and participation in board and committee meetings. Doing so also allows the director to discharge his or her legal duties, discussed below, of due care and loyalty to the company. Periodic, honest, and dynamic feedback among the board can be very helpful. This can be done informally by telephone chats among the directors one-on-one, or as a group, or on a more formal basis, although the formalistic aspects of such feedback frankly are rarely employed in practice by most boards. Formalistic aspects of board feedback on itself can include one or more of the following:

- Board members' periodic individual self-evaluation
- Periodic written self-evaluations, coupled with one-on-one peer review and mutual review, by the chairman with each board member
- Periodic meetings of or conference calls among board informally to discuss board effectiveness and specific ways to do better
- Periodic candid review by the board as a whole as to whether and how the board might best be restructured ("rump sessions" or "backroom politics" among subsets of directors on such issues need to be avoided)
- One- or two-day offsite meetings once a year with the board alone (no management) for the board to review its own performance

Each board member should ask himself or herself the following questions fairly regularly; if the answer to any of these is "no," the director should consider resigning to make room for a more productive and involved board member:

- "Have I talked with the company in the past month?"
- "Have I visited the company's Web site at least once in the past month to see what's new?"
- "Have I read the materials and information I've received on the company?"
- "Have I done anything productive for the company in the past three months, besides attending meetings?"

Directors should provide guidance to, and incisive (but not excessive or abrasive) questioning of management with respect to what is going on currently and what is predicted to take place in the future in the environment external to the company. For example, the board should determine and remain aware of what is actually happening in the health care market generally; where the FDA appears to be going on key issues, developments by both the company and its competitors in key technology areas to the extent available from public sources, generally what competitors appear to be doing, and similar appropriate "market intelligence."

Board members should also engage, to an appropriate degree, in "viral marketing" for the company and its technology. For example, in business and social settings, board members should talk, in a way that does not breach their duties of confidentiality or legal privilege, about the company and their enthusiasm for it. If the board member does not feel that way, he or she should say nothing. It is preferable to remain silent than to criticize, or even damn by faint praise, the company or any of its management, investors, or other board members in public; carried to an extreme, such behavior could constitute legally actionable disparagement, slander, or libel. Board criticism of the company, management, or each other, should be discussed in a civilized manner at, or before, board meetings, and never elsewhere.

Duties of the Board

Each member of the board has two fundamental and well-established legal duties.

First, he or she has a duty of due care, which involves:

- Getting enough information upon which to reasonably base the relevant decision,
- Taking enough time to adequately review the relevant information prior to the making of the decision, and
- Taking enough time in the decision-making process, including the board or committee meeting, to discuss the information and other relevant factors, before making the decision.

Secondly, each board member has a duty of loyalty, which involves:

- Disclosing to the board any actual or potential conflict of interest of the director, whether of a personal, or competitive business, or financial nature, with respect to matters involving the company;
- Abstaining, in most cases, on the decision if there is an actual or potential conflict, and as noted below, perhaps leaving the board meeting during the discussion of the matter, to allow the most open and productive board process for decision-making on the matter; and
- Acting in the best interests of the company and all of the company's shareholders. Obviously, this can be a problem if the director is on two boards on opposite sides of the actual or potential conflict. In such cases, the director should consider absenting himself or herself from all board discussions—likely on both sides, and both during the meeting and any informal board discussion—about any potentially relevant matters.

Effective Board/Management Communications

From a legal perspective, any director is entitled to access to anyone in the company, and to any company files or records, at any time and to any extent that is reasonable. This policy exists to allow the director to provide proper fiduciary oversight of management (not to provide management of the company, which is the job of management itself).

Normally, however, a director should not call directly and discuss company affairs with officers other than the CEO unless the director has a concern about the performance of the CEO and/or about the quality or completeness of information, such as financial figures, sales or technology development information, or any information being furnished to the board by the CEO, CFO, or other lead officer. Further, if the director suspects that something inappropriate is going on, such as books being cooked, the "shipping of bricks to the warehouse" and calling it sales, and the like, the director has the duty to inquire as deeply as he or she believes necessary if the director believes, in good faith, that there is a need to do so in order to properly discharge his or her fiduciary duties to the company and its shareholders.

However, if a director goes too far in such inquiry, without apparent and valid reason, the director may lose some of the protection of the business judgment rule, which is a well-established legal principle pursuant to which courts generally are loathe to second-guess or overturn decisions by directors unless there is evidence that the directors did not discharge their duties of due care and loyalty. Courts will interpret this precept more or less loosely according to the court's view of the egregiousness of the board's behavior and the way in which the process of decision, or lack thereof, by the board was conducted. For example, directors are entitled to rely upon reports, both written and oral, to them from management and from experts, including reports and advice from counsel and accountants, unless the directors know, or have reason to suspect, that such reports or advice may be incorrect or misleading (in which case, the directors have a duty to inquire further until they are satisfied that they have an accurate and complete set of facts upon which to act).

In keeping the board aware of relevant information, it is not productive for management to spring financials, particularly in unreadable tiny-type overheads, on the directors for the first time at the meeting. Consequently, board books (i.e., background materials for the meeting), detailed as appropriate for the issues to be discussed, should be made available to the directors a reasonable period of time before the meeting, typically about a week or less, but not the day before, or of, the meeting absent a compelling late-breaking reason to update prior-supplied board book information. Board books should contain all relevant information without overloading directors. Unfortunately, board books have a way of winding up in directors' files, outside of the company, which may be dangerous if any litigation occurs. To mitigate this occurrence, some companies collect board books back from the directors at the end of each meeting, particularly in public companies.

Some companies circulate board materials in two separate books. The first, which is for the general session, circulated to the board and to key management members, contains relevant general session presentation materials on such matters as technology development, business development, finance, and other business matters, but not containing prior meeting minutes, nor stock grants, nor any sensitive personnel or competitive issues. Board members then also receive a separate package, which goes only to the board members and (if they have the contract right as part of their investment documentation, and under a condition of confidentiality in any event) to investor board observers, with minutes and stock grant and other sensitive information.

Privately held companies tend to break board meetings into two sessions. During the first session, senior management is present, and the management presentations are given. In the second session, all management, usually including the CFO but not the CEO if he or she also is a director, will leave so the board can meet in true executive session to deal with relevant matters, including stock grants, personnel matters, and other sensitive issues. The issue of whether to do the executive session at the beginning or the end is up to each board, but most boards find that doing it at the end affords the freedom of time to discuss sensitive matters as long as necessary without being pressured by the fact that "management is waiting at the door to report." It is generally a good idea to establish the two-session protocol, and to establish it early in the company's history.

Some companies do allow senior management to sit in during some or all of the board meeting, but if management, apart from the CEO and occasionally the CFO, is to conduct a presentation during some part of the meeting, it is generally preferable to have the relevant members of management come in individually at the time of their presentation, then leave immediately following their presentation. Many privately held companies, however, allow the CFO to be present during the entire board meeting, or at least during the entire general session.

Major issues need to be run by the board by the CEO ahead of the board meeting (typically at a board and CEO dinner the night before the board meeting) and blessed or modified as the board members deem appropriate; this helps prevent the board meeting from becoming a "real-time wrestling ring." Arriving at a shared, or at least validated, decision before the meeting helps emphasize the team mentality of the board, as discussed earlier, and presents a more effective and professional appearance within the board meeting, especially as to matters discussed with management apart from the

CEO present (it is not productive to have "the parents fighting in front of the kids" in a board meeting).

Most boards appreciate not only what the CEO knows, but how fast the CEO can learn and react properly to changing situations. The ability to handle and communicate negative news with respect to the company is an excellent metric of a CEO's merit. If there is bad news:

- The board needs to hear about it from the CEO first, rather than from someone else; surprise is not the CEO's or the board's friend.
- It is better for the CEO to have twelve real problems to present when there are only six that truly matter, than to present three problems when there are in fact six that really matter. Presenting a reasonable range of real problems allows the CEO to keep the board informed as to the breadth of the problems, while focusing on the priority problems, rather than under-informing the board as to lesser important, but still important problems. Also, a board may view a given presented issue as more important than the CEO has presented it.
- The problems should be presented with the true flavor of their immediate or long-term seriousness, and with solutions in hand; alternatively, they should be presented with the caveat, if true, that the CEO has only recently become aware of the problem(s) and is actively reviewing the facts.

Finally, it is as detrimental for a CEO to listen to everything a board says as it is for the CEO to listen to nothing the board says.

The Simple Key: Working Together for Success

The success of a board, and in turn the company itself, stems from a teamwork approach focused on informed, dynamic, and open communication among the board, and between the board and the CEO, as well as mutual respect and a commitment to proactive, appropriate, and helpful involvement by the board and the CEO in ensuring that these precepts are part of the company's—and the board's—core values and culture.

Bruce W. Jenett practices corporate law and is co-leader of Heller Ehrman's life sciences practice. Mr. Jenett's practice is focused on the representation of domestic and international high-technology business clients, primarily in the life sciences industry. He is highly experienced in equity and debt financing, licensing and distribution, strategic alliances, joint ventures, and mergers and acquisitions, as well as general counseling issues, representing both startups and large multinational corporations. Mr. Jenett has represented such clients as Bayer Corporation and Bayer AG, the Biotechnology Industry Organization, CardioGenesis Corporation, Calypte Biomedical Corporation, Indigo Medical, Inc., Gryphon Therapeutics, Hitachi Chemical Diagnostics, Entelos, Inc., Nautilus Biotech S.A., Pelikan Technologies, Inc., Thoratec Corporation, TransMolecular, Inc., VisionCare Ophthalmic Technologies, Inc., Astellas (Yamanouchi) Venture Capital LLC, and numerous life sciences and telecommunications startup incubators.

Mr. Jenett is a frequent speaker on corporate and finance issues to both business and attorney audiences. He is a member of the Bioethics Committee of the Biotechnology Industry Organization, the advisory board of the journal Silicon Valley Business Ink, *the BayBio board of directors, and San Francisco Mayor Gavin Newsom's Biotech Advisory Council. Mr. Jenett also is named by Chambers & Partners as one of America's leading business lawyers.*

Mr. Jenett received his B.A. from Princeton University (1969) and his law degree from Georgetown University Law Center (1976).

Guidelines to Follow for FDA Approval

Gary L. Yingling
Partner
Kirkpatrick & Lockhart Nicholson Graham LLP

Working with a Client to Determine Legal Exposure Risk

When a client comes to us with biotechnology questions, first we need to discuss the client's intentions for the product and what market they want to be in. Once we know this, we can start down the path into greater detail about the product and the regulatory and legal issues they will face.

Biotechnology can involve a drug, a biologic, a medical device, or food, so we must figure out how the client's biotechnology product legally fits into the United States Food and Drug Administration's (FDA) regulatory system, or whether the product is regulated by another federal agency, such as the United States Department of Agriculture (USDA) or the Environmental Protection Agency (EPA). What kind of review or approval process is necessary? Under the FDA's regulations in the food area, there may be some biotechnology changes that do not require any kind of formal, or even informal, submission to the FDA. However, in the pharmaceutical and biological areas, any biotechnology change will require some kind of FDA approval. The question then becomes, how detailed is the approval process, and what needs to be done to complete it?

The most difficult approval process involves a pharmaceutical product with a biotechnology change. This pharmaceutical product can be either a biologic or what is called a drug. Most likely, this type of change will require a new drug application (NDA) (21 U.S.C. § 505(b)). For a biological, if the change is well characterized, it will be regulated as a new drug application.

What does the biotechnology product do? What is its advantage compared to products currently in the marketplace? How will the product be used? If it's a drug, what is the disease or medical condition for which it will be used?

If a client comes to us claiming to have a pharmaceutical biotechnology advantage in treating a disease, then we must assume the client has some kind of animal research to support the claim. Prior to performing human testing, one must submit an investigational new drug application (IND) for human research to the FDA (21 U.S.C. § 505(i)). The IND submission requires a certain amount of data gathering, including manufacturing information on the product, as well as animal studies that have been conducted. The IND must have a protocol for the clinical research, such as: What kind of study will be completed? How many patients will be in the study? What will be the end point in relationship to the diseased condition? Will the product improve a patient's quality of life or cure a specific disease like lymphoma?

There are really three stages to the IND process. The first phase of the IND is safety. The second phase shows how the drug works and determines dosage levels. The third phase is larger and would prove that the drug works.

The sponsor of the drug is required to pay a user fee, which this year is $600,000, at the time of the new drug application (NDA) filing. The studies will evolve over time. Phase I, Phase II, and Phase III will probably take anywhere from four to eight years to gather data for the application filing. The NDA is filed when all of the clinical and animal data to support approval of the application are collected. Then the application review and approval will probably take a year.

The approval process for a medical device may be shorter, depending on whether there is already a "510(k)" device in the marketplace with biotechnology innovations. If that's the case, one can market a similar device with a 510(k) submission. If one is submitting a new biotechnology use for the device, then you must file a pre-market approval (PMA). This submission looks more like a new drug application as to the data requirements, and requires an investigational device exemption (IDE) filing.

If the biotechnology is in the food area, the FDA doesn't regulate it in the same way. However, there are some rules about what the agency expects in relationship to biotechnology and food. The FDA has "monitored" but not really regulated many biotechnology changes through the generally recognized as safe (GRAS) notification system, which is basically a notice and comment procedure. The firm files a notice, and the FDA comments. In certain cases, like the biotech tomato, the FDA reviews as a food additive. One question may be whether it involves field crops, which would mean USDA involvement. There are very specific USDA rules about biotechnology changes in relationship to crops and doing studies.

Facing Legal Issues and the Approval Process

There aren't necessarily legal issues that are more important to a biotechnology company, nor do they face a separate set of legal rules. The question is: What is the use of the company's proposed product? Once one knows that, one looks at the regulatory systems that the FDA or the USDA (on the agriculture side) has in place in terms of the regulation. There is not a separate statutory section for biotechnology.

One also needs to determine the biotechnology research's level of novelty. How far is the firm asking the FDA to stretch its thinking? It is a regulatory agency. As such, it is very comfortable doing what it did yesterday, the day before, or the day before that. When one goes to the FDA and asks them to think outside the box and consider something that is really different, their response is the same: File the IND or the IDE for testing and submit either an NDA or a PMA application for approval. One must then be able to answer the FDA's questions relating to the approval; they will ask science and toxicity questions, and they will want to know the drug or device's affect on the human body and if it has any kind of genetic effect. All of those kinds of questions may be new and may be different. The biotechnology company may find itself walking the agency through the process, taking the agency by the hand and explaining what the drug or device is, how it works, and what is known about it. If my client has a pharmaceutical or device product, I would recommend a pre-IND or IDE meeting with the FDA. They should contact the agency and provide background information on their product plan. This way, when the application is filed, it's not a complete surprise to the agency.

FDA reviewers work in very specific areas, so one may find themselves presenting the agency with new science. When this happens, one needs to realize that the agency needs to get comfortable with the new science because the reviewer's obligation is to tell the FDA and all U.S. citizens, "We approve this drug or device, and we think it is safe and effective. We think that if your doctor prescribes this and you follow the directions, this drug or device will help you. It may have side effects for some people, but on a risk-benefit ratio, this drug or device has a benefit for you."

If I'm a reviewer with that responsibility, I want to exercise it carefully. If you come to me with science I haven't seen, I'm going to say, "Wait a minute, stop. I've done chemistry, and I've looked at other products. This is different. You're proposing to treat the disease differently or offer a different delivery system." As a reviewer, I want to be comfortable so that when I go home at night, I'm not concerned that there is someone using this drug or device who might be harmed because I didn't ask enough questions. It's an obligation for the biotechnology company to provide as many possible answers and as much information about the "science" up front.

Biotechnology firms need to keep in mind that the FDA is a scientifically-based agency. Does it have political overtones? Absolutely. However, the bottom line is the science, and if the science is there, then the company must focus on that and make it work to their benefit. However, the other side of that coin is that one does not go in and "bare your soul." One needs to be an advocate in relationship to what the product is and the fact that there are data to support that it works. Obviously, one can't hide any data, but one wants to take the data and present it in the best light possible.

The most important thing for a biotechnology company to do is plan its research and development program, and recognize that part of that process may be educating the agency. It's also very important to bring in outside experts to the regulating agency. It is much more comfortable when a company comes in with an academic expert who says, "I've looked at this, and I think this is really neat and it works." The agency is of the belief that the company's submission has a more balanced view if a university professor is interested in the product and states that it works.

Firms need to think about these things in relationship to their application. They also need to think about the quality of the application, such as organization, pagination, and proofreading. These may seem like little things that shouldn't be a big deal, but in many respects, they are. Since the agency is not quick to respond, those little things turn out to be more significant; each day the agency spends with the application because of technical problems means another day that you are without approval and not selling the product.

Avoiding Financial Pitfalls on Regulatory Applications

If you attend a meeting where an FDA employee talks about PMA or NDA submissions, they will most likely discuss the quality of applications, and how easy the application is to work with. Is it an electronic application, and what are the electronic capabilities within the application? The easier it is to review and the more user-friendly it is, the faster the review will go and the fewer questions a firm will get.

Take a step back for a second and consider that I'm the reviewer of your application. Let's say I get to page fifty, and I don't understand what you've said, or it's different than page twenty-seven. I will have to send my questions to you in writing (although e-mail is sometimes used), and that could take several weeks or months due to my busy schedule and the process of getting my memo reviewed by a supervisor. Then, when you received my questions, you must go through your documents and review them to find the answers. You must respond via writing, and then send the document back to the agency. All of this takes a lot of time, and costs a lot of money.

Another potential pitfall is if the agency is going to inspect a manufacturing site for a pharmaceutical NDA or PMA, something they will do ninety times out of one hundred. If they have problems with the inspection, one will probably have to add about eighteen months to the approval process, because it takes time for one to respond to their observations and concerns. The agency then must review the response, maybe send out another inspector, and by then months have gone by. In addition, this process is very expensive. There is just nothing better than ensuring that your application is thorough, complete, accurate, and reflective of the manufacturing.

Laws Affecting Biotechnology Companies

In relationship to biotechnology companies, other than certain environmental ones, the laws are not significantly different than the ones affecting chemical companies in terms of pharmaceutical, device, and food companies. It is all the same statutes and regulations. The question for food is: Is it safe? The questions for a pharmaceutical and a medical device are: Is it safe, and is it effective? What is the product's risk-benefit ratio?

For many years, Congress reviewed the statute and made major changes effecting the FDA every twenty or thirty years. During the past fifteen years, Congress has gotten very active in tweaking the Food, Drug, and Cosmetic Act every two or three years. Starting around 1983, there is a bill almost every two or three years with some change. Many of the changes are minor, but nevertheless, a change in the statute means the FDA has to make adjustments. It takes the FDA approximately four or five years to make that adjustment. The FDA has been described as a large, slow-moving object that bleeds easily. It is a large organization with a significant bureaucratic system, and changes are not easy. When laws and rules are changed, the FDA has to rewrite its regulations and policies, which takes a number of steps.

Most of the international issues we have observed have involved patent and intellectual property questions. From a regulatory point of view, each country has its own regulatory systems. For foods, the European Union is much more concerned about biotechnology and much more interested in labeling foods. One will run into a lot more questions regarding food and biotechnology in the European Union and in other parts of the world.

There are number of small countries without significant regulatory systems, and they are often frightened by biotechnology. The idea of a genetic change in something means the product could potentially cause all kinds of problems as far as they're concerned. They don't know exactly what the problems are, but a biotechnology food product raises issues because one seldom has a clear advantage outside of yield, which consumers do not understand. With biotechnology in relationship to drugs, devices, or biologics, the argument is that the product is superior, either with a cure rate, fewer side effects, or something similar. With food, a tomato is a tomato. If it's more resistant to pesticide, maybe that helps the farmer, but for the person buying, it's still a tomato. For foods, it becomes easier for that fear to be used in political processes.

Going to Court

With a matter involving the FDA, one does not start out in court, but very occasionally one ends up there. When that happens, we have failed the client. The idea is to work with the company to take their innovative biotechnology through the regulatory process, and to have it be approved, used, or labeled with the minimum number of problems and issues. Litigation is a problem and a huge issue. One does not want to end up there.

When working with companies trying to get products into the marketplace, litigation is seldom an issue. However, there are two things to become concerned about when first meeting with clients. One is when clients offer a "magical cure" for things that don't have a "cure." If a client comes to us and says they have a cure for a disease such as AIDS, we have lots of questions, because the FDA will have a lot of questions. The second is when firms start talking, and it becomes clear that they don't have a good understanding of their technology. If the firm doesn't understand the technology, then the chances of convincing the agency that the technology is safe and reproducible diminish astronomically. If a firm comes to us with conceptual ideas, but without completed technology or animal studies, then our concern is that the firm may have come to us too soon. From a regulatory point of view, the company must have a completed action plan or recognize that they need one. FDA approvals are seldom quick and almost always different. Firms that do not recognize that will often find the process frustrating and different.

Sharing Advice

The best piece of advice we have received in terms of biotech companies is that they must be prepared to educate the reviewers as to why the biotechnology is not "different" and doesn't raise new and different issues. To prepare for this, the company must have solid science, which doesn't necessarily change from a regulatory perspective—it's the same process. It's similar to someone obtaining a driver's license at age fifteen, twenty-five, forty, seventy-five, or ninety—no matter the age, it's the same test. It's the same thing in relationship to a drug approval. Whether it's biotechnology or basic chemistry, it's still the same process.

The advice we often tell our clients is that they really need to understand their biotechnology, and they must ask questions about its safety and effectiveness. They need to know that the agency is not going to miss any questions. A company can't offer an application with something missing or unknown and think the agency won't ask about it; ninety-nine times out of one hundred, the agency will ask. Companies can't fool themselves into thinking it's not a big deal.

Changes in Laws

In the next ten years, changes in laws affecting biotechnology companies depend on where the product development goes and how fast it goes. If we look back on drug laws, the statute was significantly changed in 1938. It required that drugs be safe, but it left the FDA a great deal of flexibility. When antibiotics and insulin came in 1941, Congress basically required each batch be tested because the technology was so new. Everybody was concerned about it, and the fear was that it would get out of control. As a result, it was very tightly controlled.

If biotechnology continues to come in what I would call a slow wave, then I predict that we will not see any, or many, new laws. Instead, there will be regulations and policies that will adapt to these changes. However, if we see some significant innovation of dramatic effects or unforeseen safety issues, then there will be questions about whether the legal process was appropriately set up to deal with the technology.

While we've seen a number of significant improvements and more improvements are coming, we haven't seen the kind of leap we saw between 1938 and 1945 in terms of the availability of antibiotics. Unless biotechnology creates new wonder drugs, there probably won't be any new legislation. If it does, then there may be a whole set of information technology questions and a whole set of other issues that will come up relative to how it is regulated, controlled, and improved; all of these questions go hand in hand. Once it becomes a political issue, Congress will deal with it. If it stays a slow-rolling scientific issue, then Congress probably won't do anything, except enact laws to confirm what the agency is already doing policy-wise. While the science of biotechnology is new, the basic safety and effectiveness issues facing the FDA and the USDA have not changed, and therefore the basic regulatory process has not changed much either.

Gary L. Yingling is a partner with Kirkpatrick & Lockhart Nicholson Graham LLP. His practice focuses on regulatory and legal issues concerning food, drugs, medical devices, and cosmetics. His primary efforts have been working with the FDA, but he also represents clients before the USDA Food Safety and Inspection Service, the EPA, the CPSC, the FTC, and various states. His clients have included individuals, partnerships, and corporations, and have involved labeling, importation, regulatory (marketing) strategy, recalls, seizures, and criminal matters. His work in food has ranged from ingredient safety questions to product labeling. In the drug area, Mr. Yingling's work has ranged from preparing INDs to product labeling. An area of particular interest has been clinical research/contract research organization/sponsor matters.

Continuous Capture and Assessment of Intellectual Property

Denise M. Kettelberger
Partner
Merchant & Gould P.C.

Katherine M. Kowalchyk
Partner
Merchant & Gould P.C.

Large or small, biotech companies generate, analyze, and maintain large amounts of data. Managing this data to build an effective intellectual property portfolio that can provide commercial advantage while clearing minefields of competitor's patents takes time, money, and commitment. We recommend thoughtful and strategic management of intellectual property assets, aligning a strategic IP plan with company goals. Each of our client companies has different goals they wish to achieve, and each likewise has a different, client-centric strategic IP plan. We work with our clients to understand their goals and objectives, and to develop an IP strategy that meets their needs.

Experimental data, related information, hypotheses, correlations, logical extensions, and more generally surround a single biotech invention. Biotechnology projects can extend over long periods of time, over many scientific programs and company employees, and across management by several counsel before developing into a desirable lead product. Laboratory notebooks and company reports generally describe the details of experimental results and hold exemplary data, however, the scope of the discoveries and inventions can reach much further than the working examples. Information logically following a disclosed discovery, gathered and discussed within a research group, may never be recorded in a lab notebook.

Frequent and repetitive communications with various members of the research team helps us to understand the client's view of a particular technology package and leverage this knowledge into appropriate patent disclosures supporting patent claims of desirable scope. Discussions about competing projects, review of the relevant art, and watching literature and patents in the field keeps us involved in the product development cycle and enables us to respond quickly to changing needs in a particular area.

Project progression, for example, from genomic sequence screening through initial homology and functional assays, correlation studies and analysis of variants, agonists, and antagonists, stimulates many new discoveries and inventions. Company employees generate multiple invention disclosures and many different patent applications are submitted to worldwide patent offices. The description and scope of the technology evolves with time and understanding, and is recorded in the time line of patent applications. These applications thus contain a varied scope of disclosure, claims, and filing dates, following the specific discoveries and improvements realized as the project progresses. We proactively prosecute such families of applications in a consistent manner, using of the strengths of each application to fill a specific objective in the strategic plan. Some early applications may secure broad pioneering discoveries, while later applications may focus on securing specific characteristics that provide commercial advantage. Keeping the strategic plan in mind, we focus our attentions on using the patent portfolio to maximize the company's market position and to create roadblocks for current and potential competitors.

Regular audits of technology development and disclosures, publications, trade secrets, and patent applications, as well as an informed, consistent approach to patent filing and prosecution of patent applications can help avoid loss of potentially valuable patent assets. We view regular and direct communications with our clients to be the most important part of our practice.

Obtaining Useful Claim Scope

Biotechnology inventions are complex and multifaceted. Due to the nature of the genetic sequences that form many of these inventions, a single disclosed DNA or protein sequence can be readily altered or changed to maintain or change the molecule's disclosed function. To gain commercial advantage, the scope of patent claims must necessarily cover variations of the sequence. Timing of patent application filing as well as scope of the invention disclosure in the application, including variety of working examples greatly impacts available claim scope. Before filing a patent application, we encourage our clients to take a realistic look at the scope of patent claims that might be awarded, and what, if any steps might be taken to help gain claim scope while retaining the early disclosure of the invention.

For example, a newly discovered DNA molecule encodes a polypeptide displaying at least one, and often many biological functions. These functions can be induced by an agonist or blocked by an antagonist, depending on the desired outcome. Typical agonists include fragments of the molecule, while typical antagonists include antibodies. Although nucleic acid and amino acid sequences are unique, motifs or domains within a sequence can be common with other sequences that share specific functions. Portions of the sequence outside these domains may tolerate variation. Patent claims that include in their scope as many of these variations as possible are desirable to exclude a competitor from entering the market with a similar product.

As the study of the original molecule progresses, more and more information is obtained to confirm the initial hypotheses of biological function, structure-function relationships, tolerable variation, and conserved domains. Agonists and antagonists are produced and tested for impact. During this development time we communicate often with our clients. Timely review of the technology development with the research team enables us to expand the intellectual property portfolio with strategically filed and prosecuted patent applications, describing the invention as it unfolds, yet mindful of what that disclosure teaches to others.

With our clients, we consider several factors to determine an appropriate timing of patent applications filing. These include the amount of information known, experiments and continued work in progress, risk that competing patent applications may be filed, and timing to potential product launch.

An assessment of the amount of scientific information available in view of the desired patent claim scope is necessary to decide the timing of patent filings. As discussed above, biological molecules can be readily changed yet retain the same function. A broad patent scope covering as many variations of the biological molecule possible is important to gain competitive advantage. However, the patent claim scope must be commensurate with the amount of scientific information provided in the patent disclosure. The disclosed information ca be actual experimental examples, correlations to known data, and logical expansion of the data to reach conclusions that are more predicable than not. Expectations of claim scope supported by expansion of the working examples must be balanced against what patent examiners might consider "unpredictable" in the art of biotechnology.

To bolster the disclosure's initial working examples, we may consider with our clients if specific studies designed to demonstrate a broader claim scope might be completed prior to filing, mindful of any upcoming public disclosures or competitive risks. When additional experiments are not possible, we carefully assess what the disclosure teaches to one in the art, and craft the patent application disclosure to exemplify that teaching.

Time to product launch is also an important consideration in the timing of patent application filing. For example, when the product launch may be delayed due to regulatory processes, an early patent filing may not be wise, particularly if patent term extension may not be available. In communications with our clients, we determine a patent filing strategy that will maximize the period of market exclusivity after product launch, but balanced against risk of competitive activity.

Understanding the Competitive Landscape

Commercial development and launch of a biological product, much like traditional pharmaceutical products, is expensive and can take years to complete. Before investing the company's time and money, it is important to understand if the manufacture, sale, and use of the end product might impinge intellectual property rights of others. We recommend to our clients a strategy that includes early and frequent consideration of third party intellectual property rights and a proactive strategy to address possible problems. A variety of options can be considered when third party rights are identified, including IP acquisition or licensing, strategic partnering with the owners of the IP, and challenging the validity of the intellectual property rights of others.

Together with our clients, we develop a working strategy that keeps us informed of third party competitive activity. Regular review of competitor's literature, published patent applications, and products allows us to consider and suggest proactive strategies. Such strategies might include for example: copying claims to invoke an interference where priority and/or validity can be addressed; filing continuation applications with specific claims drawn from the client's patent disclosure that target later developed competing products; opposing competitors patents countries such as Europe, requesting reexamination of a competitor's patents; and seeking a license or partnership. These mechanisms enable proactive, well-reasoned enforcement strategies as well as maintain the integrity and strength of our client's intellectual property portfolio.

Proactive means to challenge the validity of third party patents in the United States currently include third party reexamination and interference procedures. Proposed legislation seeks to add post grant opposition to this list. A U.S. opposition procedure will provide an opportunity to challenge third party patents for a limited period of time after issue and under U.S. PTO preponderance of the evidence standard of patentability rather than under the higher standard of clear and convincing evidence imposed during infringement litigation. Grounds for seeking invalidity will likely include a broad scope of patentability issues, including novelty, obviousness, enablement, and written description.

At our firm we are preparing for this new U.S. post-grant opposition proceeding with improved systems for tracking and monitoring third party patents in areas of interest to our client's patent portfolios, and acquiring additional expertise and training in *ex parte* contested patent matters. We expect biotechnology companies to quickly adopt a new U.S. opposition procedure for many reasons, including the large number of patents in this area, close competition between companies for similar molecules, and the demonstrated use of the interference procedure to challenge the validity of biotechnology patents.

Enforcement of our clients' patent rights against third parties an important part of maintaining a competitive advantage. We help our clients to assess the strength and weaknesses target patents long before a lawsuit is initiated, including assessment of patent claims, prior art, possible defenses, and cost/benefit factors. Together with our client, our client prosecution team and firm's litigation counsel explore possible alternatives such as negotiation, arbitration, reexamination, and litigation, including litigation in alternative forums such as the International Trade Commission.

Maximizing Return on Investment

We understand that our clients are not interested in building an intellectual property portfolio that does not help them gain a good return on the large investment made in research and development. Maximal returns result from maximizing the time of market exclusivity. Building and maintaining a strategic worldwide portfolio of mixed intellectual property rights, including trademarks, copyrights, and patents covering multiple aspects of the client's products is one way we help our clients to gain market exclusivity.

We work with our clients to build a patent portfolio directed a broad range of product production and use. Variants of products, methods of making products, particularly on a large scale, methods of assaying products or monitoring product quality and methods of use are just some examples of the type of patent claims that are useful for biotechnology inventions. Patent claims that cover at least some variations of the biological molecule can exclude a competitor from entering a market with a similar product. International patent protection in the countries where the product will be produced and/or marketed is also important. We approach international patent applications carefully, however, as many countries limit the type of biological inventions that can be patented and/or have particular requirement for the phrasing of medical and agricultural patent claims.

Patent term extensions can also add time to market exclusivity. One type of extension is available for those products delayed from first market entry due to U.S. regulatory processes. We confirm with our clients what patents might be eligible for regulatory extensions, and monitor the issue dates of patents to make sure the short window of opportunity to claim such patent term extensions is not missed.

In addition, patent term adjustment is also available to compensate patentees for delay caused by the Patent Office in prosecuting a patent application to issuance. Because the number of days delay caused by the U.S, PTO is reduced by the number of days delay caused by the patent applicant, we strive to complete all communications with the U.S. PTO within statutory due dates, and avoid extending the time for reply. We calculate the amount of term extension according to statutory provisions, and carefully check the PTO's extension award for accuracy, challenging the award within the appropriate time when we believe an improper adjustment has been made. These methods of extending patent term can effectively extend market exclusivity for our clients.

We also recognize that a large IP portfolio may sometimes need culling. Regular audits of the technology portfolio are recommended to determine what IP is important to existing products, products in development, products still "hot" in the research areas of interest. These audits also determine what is no longer of interest, and not likely to return to the research and development teams.

During an IP audit, we ask our clients to consider if the intellectual property rights are adequate to protect the products as they are now in the marketplace, and as improvements are expected. Do they stop competitive products? Has the product changed to rely on other IP or is more protection needed? Looking at the return for investment, the client is able to consider where the portfolio needs additions and where culling of the portfolio is appropriate.

Licensing some of the IP portfolio may generate revenue to add to the return on investment. The regular portfolio audit can identify IP, including patent applications and issued patents that might be offered to others without adversely affecting market exclusivity. Offering licenses to competitors can create an income stream that might not otherwise be realized.

Making the technology audit a regular feature of the client's strategic plan makes the procedure less burdensome and more productive, The audit also provides our clients with a better understanding of their IP portfolios assurance that the strategic plan is helping to maximize their technology investment.

Client Communications

We believe client communications are essential to a good working relationship and effective use of outside counsel. We meet regularly with our clients make sure that their needs are being met by our client teams, to facilitate understanding of complex technology and legal issues, and to foster good will. Understanding the client's business goals and objectives and timelines is very important to developing a strategic intellectual property plan and budget.

Within our office, our client teams meet on a regular basis to discuss client projects, current art issues, and strategies that have been employed successfully in particular cases. During these client team meetings, we share best practices, client requests, newly acquired knowledge, and assess work to be done. These meetings help us to keep informed as a group to provide a consistent and efficient work product for our clients.

We communicate not only to understand the client's technology but also to discuss and understand the client's business goals. Flying from our home base in Minneapolis for quarterly, preferably monthly meetings at our client's headquarters give both our client and our client teams a better understanding of the work objectives, and creates a better working partnership.

Denise Kettelberger *is a partner with Merchant & Gould and chairs the firm's ChemBio practice group. She is a registered patent attorney with a practice in counseling biotechnology and pharmaceutical clients in strategic management of intellectual property assets, including worldwide patent portfolio management, product review, infringement/validity analysis, technology audits, and prosecution matters. Ms. Kettelberger is active in the American Intellectual Property Law Association, and served as chair of the AIPLA Emerging Technologies Committee, vice-chair of the Patent Law Committee and is currently a member of the AIPLA Board of Directors. She is a frequent speaker on Biotechnology Patent Law issues.*